Radio Systems III

A textbook covering the Level III syllabus
of the Technician Education Council

D C Green

M Tech, CEng, MIERE
Senior Lecturer in Telecommunication Engineering
Willesden College of Technology

Pitman

PITMAN PUBLISHING LIMITED
128 Long Acre, London WC2E 9AN

Associated Companies
Pitman Publishing Pty Ltd, Melbourne
Pitman Publishing New Zealand Ltd, Wellington

© D. C. Green 1979

First published in Great Britain 1979
Reprinted 1981, 1982 (twice), 1984, 1985

Printed in Great Britain at The Pitman Press, Bath

ISBN 0 273 01134 0

Contents

Preface

This book provides a comprehensive coverage of the circuits and techniques used in modern radiocommunication systems and equipments.

The Technician Education Council (TEC) scheme for the education of telecommunication technicians introduces the basic principles of radio systems at the second level and a further understanding is provided by the third level unit Radio Systems III. This book has been written to provide a complete coverage of the Radio Systems III unit.

Chapters 1 and 2 cover the principles of amplitude and frequency modulation while Chapter 3 discusses the various kinds of modulator and demodulator circuits in common use. The next chapters deal with, respectively, radio-frequency transmission lines and the four types of aerial commonly employed in modern systems, namely the Yagi, the rhombic, the log-periodic and the parabolic reflector. The propagation of radio waves is the subject of Chapter 6 and then Chapter 7 is devoted to the tuned radio-frequency power amplifier. Chapters 8 and 9 then deal with the principles and practice of modern communication transmitters and receivers operating in the h.f. and the v.h.f. bands. Many of the circuits used in radio receivers, such as amplifiers and oscillators, are discussed in the companion volume *Electronics III* and are not covered in this book. Other circuits usually found in radio receivers, such as mixers, crystal filters and squelch circuits are dealt with in Chapter 10; this chapter also discusses the various ways in which a.g.c. and a.f.c. can be applied to a radio receiver. Lastly, Chapter 11 introduces the basic principles of wideband multi-channel telephony systems operating over both land and radio systems.

This book has been written on the assumption that the reader will possess a knowledge of electronics and radio equivalent to that covered by the TEC level II units, Electronics II,

Radio Systems II and Transmission Systems II. The reader should also have studied, or be concurrently studying, the level III unit Electronics III since knowledge of the operation of the bipolar and field-effect transistors and of integrated circuits is required.

The book provides a comprehensive text on radiocommunication systems that should be eminently suitable for all non-advanced students of radio engineering.

Acknowledgement is due to the Technician Education Council for their permission to use the content of the TEC unit in the appendix to this book. The Council reserve the right to amend the content of its unit at any time.

Many worked examples are provided in the text to illustrate the principles that have been discussed and each chapter concludes with a number of short exercises and longer exercises. Many of the exercises have been taken from past City and Guilds examination papers and grateful acknowledgement of permission to do so is made to the Institute. Answers to the numerical exercises will be found at the end of the book; these answers are the sole responsibility of the author and are not necessarily endorsed by the Institute.

D.C.G.

The following abbreviations for other titles in this series are used in the text:
TSII: Transmission Systems II
RSII: Radio Systems II
EII: Electronics II
EIII: Electronics III

1 Amplitude Modulation

Introduction

Amplitude modulation of a sinusoidal carrier wave is widely used in line and radio communication systems as a means of shifting, or *translating*, a signal from one frequency band to another. Frequency translation of signals is commonly used for two reasons. Firstly, the internationally recommended bandwidth for commercial quality speech is 300–3400 Hz and this figure is very much smaller than the available bandwidth of a telephone cable. This means that a cable pair is capable of simultaneously transmitting a number of speech channels provided the channels are each positioned in a different part of the frequency spectrum of the cable. This process is known as frequency-division multiplex (f.d.m.) and is capable of providing up to 10 700 channels over a single coaxial cable pair.

Secondly, frequency translation of an audio frequency signal is also used in all kinds of radio systems. Radio signals are transmitted and received by means of aerials, but since no kind of aerial can operate at such low frequencies it is necessary to shift each signal to some higher frequency. It is, of course, necessary to carefully choose the frequency bands to which the signals are moved in order to ensure that each service within a given geographical area operates at a different frequency. In practice, the frequency bands which are used for particular purposes are allocated in accord with the recommendations of the International Telecommunication Union (I.T.U.).

Principles of Amplitude Modulation

To obtain the maximum utilization of an available frequency spectrum, it is necessary for signals to be frequency translated to occupy different parts of that frequency spectrum. In many of the systems to be described later in this book, frequency translation of a signal is accomplished by the signal amplitude modulating a carrier of appropriate frequency.

The general expression for a sinusoidal carrier wave is

$$v = V_c \sin(\omega_c t + \theta) \tag{1.1}$$

where v is the instantaneous carrier voltage, V_c is the peak value, or amplitude, of the carrier voltage, ω_c is 2π times the carrier frequency, and θ is the phase of the carrier voltage at time $t = 0$. Here, θ will be taken as being equal to zero.

For the carrier wave to be amplitude modulated, the amplitude of the carrier voltage must be varied in accordance with the characteristics of the modulating signal. Suppose the modulating signal is sinusoidal and is given by $v = V_m \sin \omega_m t$, where V_m is its peak value and ω_m is 2π times its frequency. The amplitude of the carrier must then vary sinusoidally about a mean value of V_c volts. The peak value of this variation should be V_m volts, and the frequency of the variation should be $\omega_m/2\pi$ hertz. The amplitude of the modulated carrier wave is therefore $V_c + V_m \sin \omega_m t$, and the expression for the instantaneous voltage of an amplitude-modulated wave is

$$v = (V_c + V_m \sin \omega_m t) \sin \omega_c t \tag{1.2}$$

Multiplying out,

$$v = V_c \sin \omega_c t + V_m \sin \omega_m t \sin \omega_c t \tag{1.3}$$

Using the trigonometric identity

$$2 \sin A \sin B = \cos(A - B) - \cos(A + B)$$

equation (1.3) may be rewritten as

$$v = V_c \sin \omega_c t + \frac{V_m}{2} \cos(\omega_c - \omega_m)t - \frac{V_m}{2} \cos(\omega_c + \omega_m)t \tag{1.4}$$

This equation shows that a sinusoidally modulated carrier wave contains components at three different frequencies:

the original carrier frequency, $f_c = \omega_c/2\pi$
the *lower sidefrequency*, $f_c - f_m = (\omega_c - \omega_m)/2\pi$
the *upper sidefrequency*, $f_c + f_m = (\omega_c + \omega_m)/2\pi$

The modulating signal frequency f_m is *not* present.

The maximum amplitude of the modulated wave occurs when $\sin \omega_m t = 1$, and is $V_c + V_m$.

The minimum amplitude occurs when $\sin \omega_m t = -1$, and is $V_c - V_m$.

Fig. 1.1 shows the waveform of a sinusoidally modulated wave, the outline of which is known as the *modulation envelope*. In practice, the modulating signal is rarely sinusoidal; when this is the case, each component frequency of the modulating signal produces corresponding upper and lower sidefrequencies in the modulated wave, and the modulation envelope

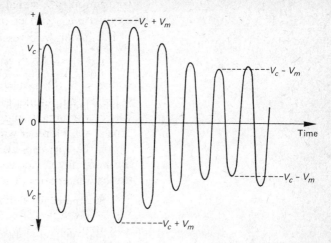

Fig. 1.1   Amplitude modulated wave

has the same waveform as the modulating signal. The band of sidefrequencies below the carrier frequency is known as the *lower sideband*, and the band above the carrier forms the *upper sideband*.

EXAMPLE 1.1

A 4 MHz carrier wave is amplitude modulated by the band of audiofrequencies 300–3400 Hz. Determine (a) the frequencies contained in the modulated wave and (b) the bandwidth occupied by the signal.

Solution
(a) Using equation (1.4) the components of the modulated wave are
  (i) The carrier frequency $f_c = 4$ MHz,
  (ii) The lower sideband frequencies 4 MHz − (300–3400) Hz or 3 996 600 Hz to 3 999 700 Hz   (*Ans.*)
  (iii) The upper sideband frequencies 4 MHz + (300–3400) Hz or 4 000 300 Hz to 4 003 400 Hz.   (*Ans.*)
(b) The necessary bandwidth = highest frequency − lowest frequency
$$= 4\,003\,400 - 3\,996\,600$$
$$= 6800 \text{ Hz}   (Ans.)$$
Note that the bandwidth occupied by the modulated wave is equal to twice the highest frequency contained in the modulating signal. This is always the case when the carrier frequency is higher than the highest modulating frequency.

EXAMPLE 1.2

A carrier wave of frequency 1 MHz and amplitude 10 V is amplitude modulated by a sinusoidal modulating signal. If the lower sidefrequency is 999 kHz and its voltage is 20 dB below the carrier amplitude, calculate the amplitude and frequency of the modulating signal.

Solution
Since the carrier and lower sidefrequency voltages are developed across the same impedances, the expression

$$\text{Loss} = 20 \log_{10} (V_1/V_2) \text{ decibels}$$

can be used.
Thus

$$20 = 20 \log_{10} \left(\frac{\text{Carrier voltage}}{\text{Sidefrequency voltage}} \right)$$

Taking antilog$_{10}$ of both sides,

$$10 = \left(\frac{\text{Carrier voltage}}{\text{Sidefrequency voltage}} \right)$$

so that

$$\text{Sidefrequency voltage} = 1 \text{ V}$$

From equation (1.4), the amplitude of a sidefrequency is equal to one-half of the voltage of the modulating signal; therefore

$$\text{Modulating signal amplitude} = 2 \text{ V} \quad (Ans.)$$

The lower sidefrequency is equal to the carrier frequency f_c minus the modulating frequency f_m, i.e. $f_c - f_m$ so that

$$f_m = 1000 - 999 = 1 \text{ kHz} \quad (Ans.)$$

There are two ways in which the frequency components of an amplitude-modulated wave may be represented by a diagram:

(1) Each component can be shown by an arrow that is drawn perpendicularly to the frequency axis as shown by Fig. 1.2a; it has been assumed that the carrier wave, at frequency f_c, has been modulated by a signal containing two components at frequencies f_1 and f_2. The lengths of the arrows are drawn in proportion to the AMPLITUDES of the components they each represent. This method of representing an a.m. wave is satisfactory when only a few components are involved but it rapidly becomes impractical when speech signals are involved.

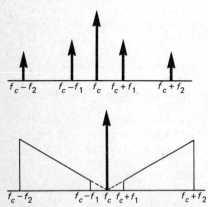

Fig. 1.2 (a) The frequency spectrum of an amplitude-modulated wave. (b) The method of representing the sidebands of amplitude modulation

(2) The method of representation usually employed, particularly in connection with multi-channel telephony systems, is shown in Fig. 1.2b. The lower and upper sidebands are each represented by a truncated triangle, in which the vertical ordinates are made proportional to the modulating FREQUENCY and no account is taken of amplitude. The upper sideband is said to be erect because its highest sidefrequency, $f_c + f_2$, corresponds to the highest frequency f_2 in the modulating signal. Conversely, the lower sideband is said to be inverted because its highest frequency component, $f_c - f_1$, is produced by the lowest modulating frequency f_1.

Modulation Factor

The modulation factor m of an amplitude-modulated wave is given by

$$m = \frac{\text{Maximum amplitude} - \text{Minimum amplitude}}{\text{Maximum amplitude} + \text{Minimum amplitude}} \qquad (1.5)$$

When expressed as a percentage, m is known as the percentage modulation, or the DEPTH OF MODULATION. For sinusoidal modulation the maximum amplitude of the modulation envelope is, from equation (1.2), $V_c + V_m$ and the minimum amplitude is $V_c - V_m$, Hence

$$m = \frac{(V_c + V_m) - (V_c - V_m)}{(V_c + V_m) + (V_c - V_m)} = \frac{V_m}{V_c} \qquad (1.6)$$

EXAMPLE 1.3

Draw the waveform of a carrier wave which has been sinusoidally amplitude modulated to a depth of 25%. If the amplitude of the unmodulated carrier wave is 100 V determine (a) the modulating signal voltage, (b) the amplitude of the lower side frequency component.

Solution
The maximum voltage of the modulated wave is $V_c + V_m$ and $V_m = m V_c$. Hence the maximum voltage is

$$100(1 + 0.25) = 125 \text{ V}$$

The minimum voltage of the wave is

$$V_c(1 - m) = 75 \text{ V}$$

The required amplitude-modulated waveform is shown in Fig. 1.3

(a) $V_m = m V_c = 0.25 \times 100 = 25 \text{ V}$ (*Ans.*)

(b) $V_{LSF} = m V_c / 2 = 25/2 = 12.5 \text{ V}$ (*Ans.*)

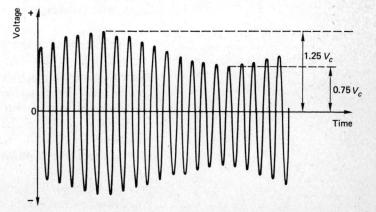

Fig. 1.3 Amplitude-modulated wave of modulation depth 25%

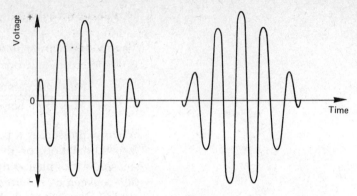

Fig. 1.4 Amplitude-modulated wave of modulation depth greater than 100%

The maximum value of the modulation factor is limited to 1 since this gives a minimum value to the envelope of $V_c(1-1)$ or zero. If a greater value of modulation factor is used, the envelope will no longer be sinusoidal (Fig. 1.4) and the waveform will contain a number of extra, unwanted frequency components.

Power Contained in an Amplitude-modulated Wave

The expression for the instantaneous voltage of an amplitude-modulated wave, equation (1.2), can be rewritten in terms of the modulation factor m:

$$v = V_c\left(1 + \frac{V_m}{V_c}\sin \omega_m t\right)\sin \omega_c t \qquad (1.7)$$

$$= V_c(1 + m\sin \omega_m t)\sin \omega_c t$$

$$= V_c \sin \omega_c t + \tfrac{1}{2}mV_c[\cos (\omega_c - \omega_m)t - \cos (\omega_c + \omega_m)t]$$

$$(1.8)$$

The power developed by an amplitude-modulated wave is the sum of the powers developed by the carrier frequency, upper sidefrequency and lower sidefrequency components. The carrier power is

$$\left(\frac{V_c}{\sqrt{2}}\right)^2 \frac{1}{R} \quad \text{or} \quad \frac{V_c^2}{2R}\,\text{watts}$$

and the power developed by each of the two sidefrequencies is

$$\left(\frac{mV_c}{2\sqrt{2}}\right)^2 \frac{1}{R} \quad \text{or} \quad \frac{m^2 V_c^2}{8R}\,\text{watts}$$

so that the TOTAL POWER is

$$P_t = \frac{V_c^2}{2R} + \frac{m^2 V_c^2}{8R} + \frac{m^2 V_c^2}{8R} = \frac{V_c^2}{2R}\left(1 + \frac{m^2}{2}\right) = P_c\left(1 + \frac{m^2}{2}\right)\text{watts} \quad (1.9)$$

As previously mentioned, the maximum modulation factor used in practice is $m = 1$, and for this condition P_t is one and a half times the carrier power. For maximum modulation conditions, therefore, only one-third of the total power is contained in the sidefrequencies. Since it is the sidefrequencies that carry the intelligence, amplitude modulation is not a very efficient system when considered on a power basis.

EXAMPLE 1.4

The power dissipated by an amplitude-modulated wave is 100 W when its depth of modulation is 40%. What modulation depth m is necessary to increase the power to 120 W?

Solution
From equation (1.9),

$$100 = P_c \left(1 + \frac{0.4^2}{2} \right) \quad \text{or} \quad P_c = \frac{100}{1.08} \text{ watts}$$

When the depth of modulation is altered to m, the total power increases to 120 W. Therefore

$$120 = \frac{100}{1.08} (1 + \tfrac{1}{2}m^2)$$

$$\frac{120 \times 1.08}{100} = 1 + \tfrac{1}{2}m^2$$

$$\tfrac{1}{2}m^2 = 1.2 \times 1.08 - 1$$

$$m = \sqrt{0.592} = 0.769 \qquad (Ans.)$$

EXAMPLE 1.5

A 1 kW carrier is amplitude modulated by a sinusoidal signal to a depth of 50%. Calculate the power at the lower sidefrequency and determine what percentage it is of the total power.

Solution
From equation (1.9)

$$P_t = 1000(1 + \tfrac{1}{2}0.5^2) = 1000 + 125$$

The carrier power is 1000 W so clearly the total sidefrequency power is 125 W. The amplitudes of the two sidefrequencies are equal and so the sidefrequencies will dissipate equal powers. Therefore

$$\text{Lower sidefrequency power} = \frac{125}{2} = 62.5 \text{ W} \qquad (Ans.)$$

The total power is 1125 W, hence the lower sidefrequency power expressed as a percentage of the total power is

$$\frac{62.5}{1125} \times 100 \quad \text{or} \quad 5.56\% \qquad (Ans.)$$

R.M.S. Value of an Amplitude-modulated Wave

If the r.m.s. voltage of an amplitude-modulated wave is V, then the power P_t dissipated by that wave in a resistance R is given by

$$P_t = \frac{V^2}{R} = P_c(1 + \tfrac{1}{2}m^2) \text{ W}$$

The power dissipated by the carrier component alone is

$$P_c = \frac{V_c^2}{2R} \text{ W}$$

Therefore

$$\frac{P_t}{P_c} = \frac{2V^2}{V_c^2} = \frac{P_c(1 + \tfrac{1}{2}m^2)}{P_c}$$

$$2V^2 = V_c^2(1 + \tfrac{1}{2}m^2)$$

$$V = \frac{V_c}{\sqrt{2}}\sqrt{(1 + \tfrac{1}{2}m^2)} \tag{1.10}$$

EXAMPLE 1.6

The r.m.s. value of the current flowing in an aerial is 50 A when the current is unmodulated. When the current is sinusoidally modulated, the output current rises to 56 A. Determine the depth of modulation of the current waveform.

Solution
From equation (1.10)

$$56 = 50\sqrt{(1 + \tfrac{1}{2}m^2)}$$

$$\left(\frac{56}{50}\right)^2 = 1 + \tfrac{1}{2}m^2$$

$$m = \sqrt{2\left[\left(\frac{56}{50}\right)^2 - 1\right]} = 0.713 \quad (Ans.)$$

The double-sideband full-carrier (d.s.b.) system of amplitude modulation can be demodulated by a relatively simple circuit which responds to the variations of the envelope of the wave. Mainly for this reason the d.s.b. system is used for sound broadcasting in the long and medium wavebands. The disadvantage of d.s.b. working, made apparent by Example 1.5, is that the greater part of the transmitted power is associated with the carrier component and this carries no information. Many radio-telephony systems use a more efficient method of amplitude modulation.

Double-sideband Suppressed-carrier Amplitude Modulation

The majority of the power contained in an amplitude-modulated wave is developed by the carrier component. Since this component carries no information, it may be suppressed during the modulation process by means of a BALANCED MODULATOR. All the transmitted power is then associated with the upper and lower sidebands.

The waveform of a double-sideband suppressed-carrier (d.s.b.s.c.) voltage is shown in Fig. 1.5. Fig. 1.5 has been drawn on the assumption that a 10 kHz carrier wave is amplitude modulated by a 1 kHz sinusoidal wave to produce lower and upper sidefrequencies of 9 kHz and 11 kHz respectively. With the carrier component suppressed, the d.s.b.s.c. waveform is the resultant of the 9 kHz and 11 kHz waveforms (Figs 1.5*a* and *b*) and is shown in Fig. 1.5*c*. The envelope of the resultant waveform is not sinusoidal and this is an indication that a

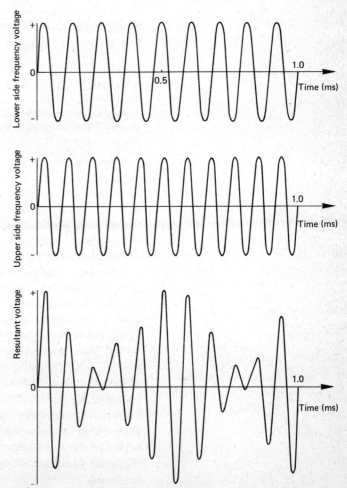

Fig. 1.5 The formation of a d.s.b.s.c. wave by adding the components at the lower and upper sidefrequencies

d.s.b.s.c. signal cannot be demodulated with the simple envelope detector which is available for d.s.b. full carrier demodulation.

For demodulation to be achieved, it is necessary for the carrier component to be re-inserted at the receiver with *both* the correct frequency *and* phase. The first of these requirements can be satisfied if the receiver circuitry includes an oscillator of sufficiently high stability, such as a crystal oscillator. The second requirement is much more difficult to satisfy and led to the rejection of this version of amplitude modulation in the past. Nowadays, modern developments, particularly in the field of integrated circuits, have considerably reduced these difficulties, and d.s.b.s.c. finds an application in two particular systems. These are the transmission of the colour information in the colour television system of the U.K., and the transmission of the stereo information in v.h.f. frequency-modulated sound broadcast signals.

Single-sideband Suppressed-carrier Amplitude Modulation

The information represented by the modulating signal is contained in both the upper and the lower sidebands, since each modulating frequency f_1 produces corresponding upper and lower sidefrequencies $f_c \pm f_1$. It is therefore unnecessary to transmit both sidebands; either sideband can be suppressed at the transmitter without any loss of information.

When the modulating signal is of sinusoidal waveform, the transmitted sidefrequency will be a sine wave of constant amplitude. Should this signal be applied to an envelope d.s.b. detector, a direct voltage output would be obtained and not the original modulating signal. This means that, once again, demodulation using an envelope detector is not possible. For demodulation to be achieved, the carrier component must be re-inserted at the correct frequency. Now, however, the phase of the re-inserted carrier does not matter and the design of the receiver is considerably eased. This method of operation is known as single-sideband suppressed-carrier (s.s.b.s.c.) amplitude modulation, frequently known simply as s.s.b.

The basic principle of operation of an s.s.b. system is shown in Fig. 1.6. The modulating signal is applied to a balanced modulator along with the carrier wave generated by a high-stability oscillator. The output of the balanced modulator consists of the upper and lower sidebands only. The carrier component is *not* present having been suppressed by the action of the modulator. The d.s.b.s.c. signal is then applied to the band-pass filter whose function is to remove the unwanted sideband.

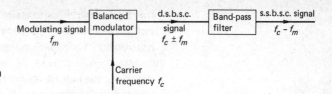

Fig. 1.6 The production of an s.s.b.s.c. signal

Single Sideband compared with Double Sideband

Single-sideband operation of a radio system has a number of advantages over double-sideband working. These advantages are as follows:

(a) The bandwidth required for an s.s.b. transmission is one half the bandwidth that must be provided for a d.s.b. signal carrying the same information. The reduced bandwidth per channel allows a greater number of channels to be provided by the transmission medium.

(b) The signal-to-noise ratio at the output of an s.s.b system is higher than at the output of the equivalent d.s.b. system. The improvement in signal-to-noise ratio has a minimum value of 9 dB when the depth of modulation is 100% and becomes larger as the depth of modulation is reduced. Exactly 3 dB of this improvement comes about because the necessary bandwidth is reduced by half, and noise power is proportional to bandwidth. The rest of the improvement arises as a result of an increase in the ratio sideband-power/total-power.

(c) A d.s.b. transmitter produces a power output at all times whereas an s.s.b. transmitter does not. This results in an increase in the overall efficiency of the transmitter.

(d) Selective fading of d.s.b. radio waves may cause considerable distortion because the carrier component may fade below the sideband level. This allows the sidebands to beat together and generate a large number of unwanted frequencies. This type of distortion does not occur in an s.s.b. system since the signal is demodulated against a locally generated carrier of constant amplitude.

(e) In a multi-channel telephony system, any non-linearity generates intermodulation products, many of which would lead to inter-channel crosstalk. The most likely sources of non-linearity distortion are the output stages of amplifiers since these are expected to handle the largest amplitude signals. Suppression of the carrier component reduces the amplitude of the signals that are applied to the output stages and in so doing minimizes the effect of non-linearity.

The main disadvantage of s.s.b. working is the need for the carrier to be re-inserted at the receiver before demodulation can take place. This requirement increases the complexity, and therefore, the cost of the radio receiver. It is for this reason that sound broadcast systems do not use single-sideband modulation.

The frequency of the re-inserted carrier must be extremely accurate if distortion of the demodulated signal is to be avoided. For speech circuits an accuracy of perhaps ±20 Hz may be adequate but for telegraphy and data signals ±2 Hz accuracy is needed. With modern receivers, the re-inserted carrier is generated by a FREQUENCY-SYNTHESIS equipment of high frequency accuracy and stability. The necessary re-insert carrier frequency accuracy is easily achieved. In older equipments a low-level *pilot carrier* is transmitted along with the wanted sideband. The pilot carrier has an amplitude of about −16 dB relative to the transmitted sideband, and is used in the receiver to operate *automatic frequency control* (a.f.c.) circuitry. The a.f.c. circuitry acts to maintain the frequency of the re-inserted carrier within the prescribed limits.

The output power of an s.s.b. transmitter is usually specified in terms of the PEAK ENVELOPE POWER (p.e.p.). The p.e.p. is the power which would be developed by a carrier whose amplitude is equal to the peak amplitude of the pilot carrier and the transmitted sideband. When the pilot carrier is not transmitted, or is neglected, the term *peak sideband* power (p.s.p.) is often used instead of p.e.p.

EXAMPLE 1.7

The output voltage of a sinusoidally modulated s.s.b. transmitter is applied across a 600 Ω resistance. If the amplitude of the transmitted sidefrequency is 60 V, calculate (*a*) the p.s.p., (*b*) the p.e.p. Assume a pilot carrier is transmitted at a level −16 dB relative to the transmitted sidefrequency.

Solution

(*a*) p.s.p. $= 60^2/600 = 6$ W   (*Ans.*)

(*b*) 16 dB $= 20 \log_{10} (60/V_{pc})$ or $V_{pc} = 9.5$ V.

Therefore, peak voltage of resultant of sidefrequency and pilot carrier is 69.5 V and

p.e.p. $= 69.5^2/600 = 8.05$ W   (*Ans.*)

Independent-sideband Amplitude Modulation

The number of s.s.b.s.c. channels which can be transmitted over a given transmission medium is determined by the minimum frequency separation of the channels. This, in turn, is set by the attenuation/frequency characteristics of the band-pass filters, since a frequency gap between adjacent channels

Fig. 1.7 The production of an i.s.b. signal

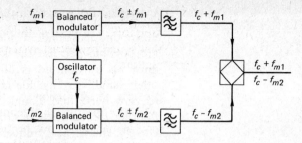

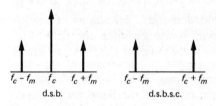

d.s.b.

d.s.b.s.c.

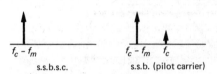

s.s.b.s.c.

s.s.b. (pilot carrier)

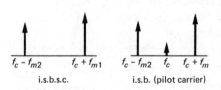

i.s.b.s.c.

i.s.b. (pilot carrier)

Fig. 1.8 The frequency spectrum diagrams of various amplitude-modulated signals

Fig. 1.9 The sidebands of various amplitude-modulated signals

must be provided in which the filter attenuation can build up. Channels can be spaced closer together, and hence further economy achieved in the utilization of the available frequency spectrum, by the use of the independent sideband (i.s.b.) system.

The basic principle of an i.s.b. system is shown by Fig. 1.7. Two modulating signals at frequencies f_{m1} and f_{m2} each modulate the same carrier frequency f_c. The outputs of the two balanced modulators are d.s.b.s.c. waveforms at frequencies $f_c \pm f_{m1}$ and $f_c \pm f_{m2}$ respectively. Two band-pass filters are used to select the upper sidefrequency $f_c + f_{m1}$ in one channel and the lower sidefrequency $f_c - f_{m2}$ in the other. The selected sidebands are combined in a hybrid coil to produce a double-sideband suppressed-carrier signal in which each sideband carries different information.

The differences between d.s.b., s.s.b. and i.s.b. can be illustrated by means of spectrum diagrams. Fig. 1.8 shows the spectrum diagrams for each type of modulation, assuming that a carrier of frequency f_c is modulated by a single frequency f_m, or in the case of i.s.b. by two sinusoidal waves at frequencies f_{m1} and f_{m2}. For a complex modulating signal, the number of arrows required is very large and an alternative kind of spectrum diagram is used in which the sidebands are represented by truncated triangles. Fig. 1.9 uses this method to illustrate the differences between the various amplitude modulation methods.

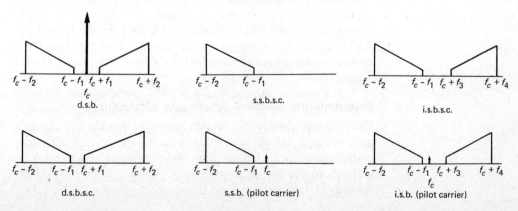

d.s.b.

s.s.b.s.c.

i.s.b.s.c.

d.s.b.s.c.

s.s.b. (pilot carrier)

i.s.b. (pilot carrier)

Vestigial-sideband Amplitude Modulation

Single and independent sideband operations of a radio system are possible because the lowest modulating frequency which must be transmitted is 300 Hz. A television signal, on the other hand, may include components at all frequencies down to zero hertz and would, in consequence, present insoluble filtering difficulties if s.s.b. operation were attempted. Conversely, the highest frequency which must be transmitted in a u.h.f. television system is 5.5 MHz and so the use of d.s.b. amplitude modulation would demand a minimum r.f. bandwidth of 11 MHz. As a compromise between the d.s.b. and s.s.b. systems of amplitude modulation, a method known as *vestigial sideband* (v.s.b.) is used.

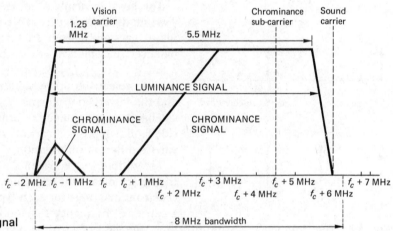

Fig. 1.10 Colour television signal

A vestigial sideband signal consists of all of the upper sideband plus a part of the lower sideband (see Fig. 1.10). The *vestige* of the lower sideband that is transmitted is not of constant maximum amplitude. The first 1.25 MHz of the lower sideband is transmitted at full amplitude and thereafter the maximum amplitude falls linearly to zero as shown. The r.f. bandwidth occupied by the v.s.b. signal is 8 MHz which is 72% of the bandwidth which would be required by the equivalent d.s.b. signal. It can be seen that the carrier of the associated sound signal is positioned 6 MHz above the vision carrier frequency.

For a colour television signal, the chrominance (colour) information is superimposed upon the luminance (monochrome) signal as shown in the figure by making the chrominance signal amplitude-modulate a 4.433 618 75 MHz *subcarrier* frequency. Quadrature double-sideband suppressed-carrier amplitude modulation is used. QUADRATURE AMPLITUDE MODULATION is a system in which two signals modulate two carriers which are at the same frequency but are 90° out of phase with one another. The two sets of sidebands produced by such a process do not become mixed up during transmission and can be separately demodulated at the receiver if two highly stable 90°-out-of-phase oscillators are available to provide the carrier re-inserts.

Measurement of Amplitude-modulated Waves

With a d.s.b. amplitude-modulated waveform, the parameter that is generally measured is the depth of modulation. This can be measured by means of a cathode oscilloscope (c.r.o.), a modulation meter, or a true r.m.s. responding ammeter or voltmeter.

(a) Use of a C.R.O.

An amplitude-modulated wave can be displayed on a c.r.o. in two different ways. The signal can be applied to the Y-input terminals and the timebase set to operate at the frequency of the modulating signal, or perhaps two or three times the modulating frequency if more than one cycle of the envelope is to be displayed. The modulation envelope is then stationary, and an amplitude-modulation envelope, such as that shown in Fig. 1.1, is displayed.

An alternative method, that makes the detection of waveform distortion easier, is to connect the modulated wave to the Y-input terminals and the modulating signal to the X-input terminals with the internal timebase switched off (see Fig. 1.11a). The resulting display is then trapezoidal, as shown at b. It can be shown that the depth of modulation of the displayed waveform is given by

$$(a-b)/(a+b) \text{ per cent}$$

The accuracy of the methods is limited mainly by the lack of discrimination that results from the need to reduce the peak-to-peak variation of the modulated wave into the area of the c.r.o. screen. The reduction in measurement accuracy is particularly noticeable when there is little difference between the maximum a and minimum b dimensions in centimetres, i.e. when the modulation factor is small.

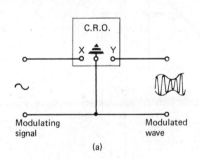

(a)

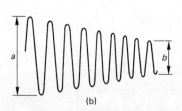

(b)

Fig. 1.11 Measurement of modulation factor using a c.r.o.

(b) Use of a Modulation Meter

A modulation meter is an instrument which has been designed for the direct measurement of modulation depth. Essentially the instrument consists of a radio receiver with a direct-coupled diode detector. If the measurement procedure specified by the manufacturer is followed carefully, accurate measurements of modulation depth can be carried out.

(c) Use of an R.M.S. Responding Ammeter

The r.m.s. value of an amplitude-modulated current wave is, from equation (1.10), given by

$$I = I_c \sqrt{(1 + \tfrac{1}{2}m^2)}$$

where I_c is the r.m.s. value of the unmodulated current waveform. The measurement procedure is as follows. The r.m.s. value of the current, with no modulation applied, is measured first. Then the modulation is applied, and the new indication of the *true* r.m.s. responding ammeter is noted. The modulation factor can be calculated using equation (1.10), or in practice, read off from a previously-calculated graph of modulation factor plotted against the ratio I_c/I.

EXAMPLE 1.8

In a measurement of modulation depth using an r.m.s. responding ammeter the unmodulated current was 50 A. Use the graph of Fig. 1.12 to determine the depth of modulation if the r.m.s. current with modulation applied is (*a*) 55 A, (*b*) 50.5 A.

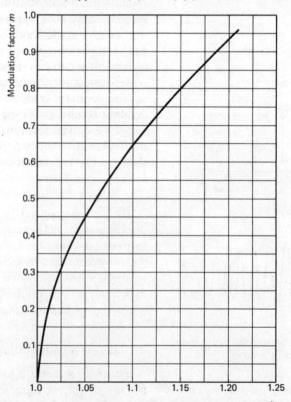

Fig. 1.12 The relationship between the ratio I/I_c of the r.m.s. currents of amplitude-modulated and unmodulated waves and the modulation factor

Solution
(a) $I/I_c = 55/50 = 1.1$

Therefore, from the graph,

   Depth of modulation $m = 65\%$ (*Ans.*)

(b) $I/I_c = 50.5/50 = 1.01$

Therefore,

   Depth of modulation $m = 14\%$ (*Ans.*)

This method of measurement is capable of accurate results for higher values of modulation depth, but for smaller values, below about 30%, the accuracy suffers because of a lack of discrimination.

In the case of a single or an independent sideband signal, the main feature of interest is the presence or absence of non-linearity distortion, since this will lead to the generation of intermodulation products with consequent inter-channel crosstalk. The usual method of measurement is to apply two audio-frequency sinusoidal waves, of equal amplitude but about 1 kHz apart in frequency, to the input terminals of the channel. The modulated output signal is then, by some means, displayed on the screen of a c.r.o. If the channel operates linearly, the two frequencies should beat together to produce the waveform shown in Fig. 1.13. Any non-linearity present in the channel equipment will manifest itself in the form of distortion of the envelope of the beat-frequency waveform.

Alternatively, if an instrument known as a *spectrum analyzer* is available, the component frequencies of the waveform can be individually displayed on the analyzer's c.r.t. screen and their amplitudes measured. Fig. 1.13*b* shows the kind of display to be expected. The required degree of linearity can then be quoted in terms of the maximum permissible amplitude of the intermodulation products.

Exercises
1.1. (*a*) What are the advantages of single sideband operation of a radio system over double sideband? (*b*) Why is the s.s.b. system not used for sound broadcasting? (*c*) Why is a pilot carrier transmitted in many s.s.b. and i.s.b. radio systems? (*d*) Why can modern systems avoid the use of a pilot carrier?
1.2. Derive an expression for the output power of a double sideband amplitude modulated transmitter in terms of the unmodulated carrier power and the depth of modulation. The output power of a radio transmitter is 1 kW when modulated to a depth of 100%. If the depth of modulation is reduced to 50% what is the power in each sideband?   (*C & G*)

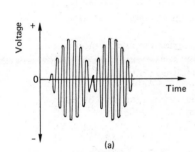
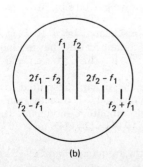

Fig. 1.13 (*a*) The signal waveform, (*b*) the spectrum diagram at the output of an i.s.b. channel whose linearity is under test

1.3. Explain how a c.r.o. can be used to measure (*a*) the modulation factor of a sinusoidally-modulated a.m. signal, (*b*) the gain of an amplifier. Discuss the likely source of inaccuracy in each method. (*C & G*)

1.4. Distinguish between the terms sidefrequency and sideband, and explain what is meant by the envelope of an amplitude-modulated waveform. Draw the waveform of a carrier wave which has been amplitude-modulated to a depth of 20% by a sinusoidal signal.

1.5. (*a*) A 50 V carrier wave of frequency 4 MHz is amplitude modulated by 10 V sinusoidal voltage. Draw the spectrum diagram of the modulated wave if the system used is (i) d.s.b., (ii) s.s.b.s.c., (iii) i.s.b.s.c.
(*b*) Repeat the above for the case when the modulating signal is the commercial speech band of frequencies 300–3400 Hz.

1.6. (*a*) Why is the use of double sideband amplitude modulation restricted to sound broadcasting?
(*b*) A 600 V carrier wave is amplitude modulated to a depth of 60%. Calculate (*a*) the modulating signal voltage, (*b*) the voltage of the lower sidefrequency.

1.7. (*a*) What do the initials (i) d.s.b., (ii) s.s.b.s.c., and (iii) i.s.b stand for?
(*b*) List the advantages of s.s.b.s.c. over d.s.b.
(*c*) What is the advantage of i.s.b. over s.s.b,?
(*d*) Explain, with the aid of a block diagram, the principle of operation of the i.s.b. system.

1.8. Explain the term *depth of modulation* as applied to an amplitude-modulated wave. Spectrum analysis of a signal shows that it comprises a carrier and one pair of sidefrequencies. Each sidefrequency voltage is 10 dB below the carrier. Calculate the depth of modulation and the total signal power if the power of the unmodulated carrier is 1 mW. (*C & G*)

1.9. (*a*) Describe a method of displaying a voltage waveform by means of an oscilloscope. Illustrate your answer by showing how the envelope of an amplitude-modulated signal is displayed.
(*b*) Draw the envelope of a carrier amplitude-modulated by a sine wave, given the following:
   Maximum peak-to-peak amplitude 80 mm
   Minimum peak-to-peak amplitude 40 mm
   Modulating frequency 2 kHz, horizontal scale 80 mm = 1 ms
Calculate the modulation factor of the signal.

Short Exercises

1.10. What is meant by each of the following: (i) d.s.b., (ii) s.s.b., (iii) d.s.b.s.c., and (iv) i.s.b.?

1.11. A 1 MHz carrier is amplitude modulated by a 10 kHz sine wave. What frequencies are contained in the modulated waveforms if the system is (i) d.s.b., (ii) d.s.b.s.c., and (iii) s.s.b.s.c.?

1.12. Draw the spectrum diagram for the signal transmitted to line for a 5-channel system in which channel 1 is directly transmitted and the remaining channels amplitude-modulate carrier frequencies of 9 kHz, 14 kHz, 19 kHz, 24 kHz and 29 kHz. Assume d.s.b. operation.

1.13. A modulating signal occupying the frequency band 68–72 kHz amplitude modulates a 100 kHz carrier. What bandwidth is occupied by the modulated wave?

1.14. A 64–68 kHz modulating signal modulates a 50 kHz carrier. Determine the bandwidth of the modulated wave.

1.15. If the power in each sidefrequency of a 25 kW carrier wave is 2 kW when the carrier is sinusoidally modulated, what is the depth of modulation?

1.16. Compare independent sideband operation of a radio system with single-sideband operation.

1.17. When a test tone is applied to one channel in an i.s.b. system, the pilot carrier is 18 dB down on the sidefrequency level. If the sidefrequency voltage is 25 V determine the voltage of the pilot carrier.

1.18. Why is the s.s.b.s.c. version of amplitude modulation not used for (*a*) sound broadcasting, (*b*) television broadcasting, (*c*) international radio-telephony (which uses i.s.b.)?

1.19. A 100 V 5 MHz carrier wave is amplitude-modulated to a depth of 60% by a 3 kHz sine wave and the carrier and upper sidefrequency components are suppressed. Draw the waveform of the transmitted signal.

1.20. Draw the waveform of d.s.b. amplitude-modulated wave for the case of a rectangular modulating signal with a mark/space ratio of 2.1.

2 Frequency Modulation

Introduction

The reasons for the use of modulation in communication systems were discussed in the previous chapter. An alternative modulation technique, known as frequency modulation, in which the modulating signal varies the *frequency* of a carrier wave, has a number of advantages over amplitude modulation. Frequency modulation is used for sound broadcasting in the v.h.f. band, for the sound signal of 625-line television broadcasting, for some mobile systems, and for multi-channel telephony systems operating in the u.h.f. band. The price which must be paid for some of the advantages of frequency modulation over d.s.b. amplitude modulation is a wider bandwidth requirement. If the bandwidth of an f.m. system is no wider than the bandwidth of the comparable d.s.b. a.m. system (narrow-band frequency modulation, n.b.f.m.), the relative merits of the two systems are not easy to determine.

Principles of Frequency Modulation

When a sinusoidal carrier wave is frequency modulated, its instantaneous frequency is caused to vary in accordance with the characteristics of the modulating signal. The modulated carrier frequency must vary either side of its nominal unmodulated frequency a number of times per second equal to the modulating frequency. The magnitude of the variation—known as the *frequency deviation*—is proportional to the amplitude of the modulating signal voltage.

The concept of frequency modulation can perhaps best be understood by considering a modulating signal of rectangular waveform, such as the waveform shown in Fig. 2.1a. Suppose the unmodulated carrier frequency is 3 MHz. The periodic time of the carrier voltage is $\frac{1}{3}$ μs and so three complete cycles

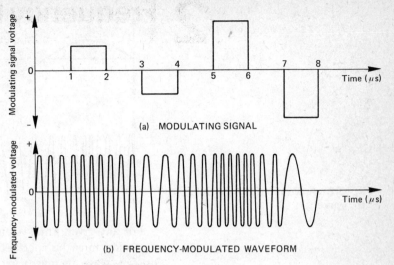

Modulating signal voltage

(a)   MODULATING SIGNAL

Frequency-modulated voltage

(b)   FREQUENCY-MODULATED WAVEFORM

Fig.   2.1   A   frequency-modulated wave

of the unmodulated carrier wave will occur in $1\,\mu$s. When, after $1\,\mu$s the voltage of the modulating signal increases to $+1$ V, the instantaneous carrier frequency increases to 4 MHz. Hence in the time interval $1\,\mu$s to $2\,\mu$s there are four complete cycles of the carrier voltage. After $2\,\mu$s the modulating signal voltage returns to 0 V and the instantaneous carrier frequency falls to its original 3 MHz. During the time interval $3\,\mu$s to $4\,\mu$s the modulating signal voltage is -1 V and the carrier frequency is reduced to 2 MHz; this means that two cycles of the carrier voltage occur in this period of time. When, after $4\,\mu$s, the modulating voltage is again 0 V, the instantaneous carrier frequency is restored to 3 MHz. At $t=5\,\mu$s the modulating voltage is $+2$ V and, since frequency deviation is proportional to signal amplitude the carrier frequency is deviated by 2 MHz to a new value of 5 MHz. Similarly, when the modulating voltage is -2 V, the deviated carrier frequency is 1 MHz. At all times the amplitude of the frequency modulated carrier wave is constant at 1 V, and this means that the modulating process does not increase the power content of the carrier wave.

When the modulating signal is of sinusoidal waveform, the frequency of the modulated carrier wave will vary sinusoidally; this is illustrated by Fig. 2.2.

Frequency Deviation

The frequency deviation of a frequency-modulated carrier wave is proportional to the amplitude of the modulating signal voltage. There is no inherent maximum value to the frequency deviation that can be obtained in a frequency-modulation system; this should be compared with amplitude modulation where the maximum amplitude deviation possible corresponds to $m=1$.

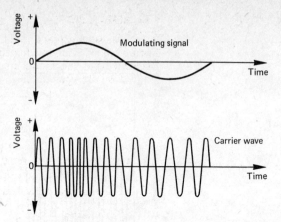

Fig. 2.2 A sinusoidally modulated f.m. wave

For any given f.m. system, a maximum allowable frequency deviation must be specified since the bandwidth occupied by an f.m. wave increases with increase in the frequency deviation. The maximum frequency deviation which is permitted to occur in a particular f.m. system is known as the RATED SYSTEM DEVIATION. Since the frequency deviation is directly proportional to the modulating signal voltage, the choice of rated system deviation sets the maximum allowable modulating signal voltage that can be applied to the frequency modulator.

EXAMPLE 2.1

A frequency-modulated system has a rated system deviation of 30 kHz. If the sensitivity of the frequency modulator is 4 kHz/V, what is the maximum allowable modulating signal voltage?

Solution

$$30 \text{ kHz} = 4 \text{ kHz/V} \times V_m$$

where V_m is the maximum allowable modulating signal voltage. Therefore,

$$V_m = \frac{30 \text{ kHz}}{4 \text{ kHz/V}} = 7.5 \text{ V} \qquad (Ans.)$$

Most of the time the amplitude of the modulating signal voltage will be less than its maximum allowable value. Then the frequency deviation of the carrier will be smaller than the rated system deviation. This can be accounted for by introducing a factor k where

$$k = \frac{\text{Modulating signal voltage}}{\text{Maximum allowable modulating signal voltage}} \qquad (2.1)$$

The frequency deviation of the carrier frequency is then given by the product kf_d, where f_d is the rated system deviation. The factor k can have any value between 0, when there is no modulating signal, and 1, when the modulating signal has its maximum permitted value.

EXAMPLE 2.2

An f.m. system has a rated system deviation of 75 kHz and this is produced by a modulating signal voltage of 10 V. Determine (*a*) the sensitivity of the modulator, and (*b*) the frequency deviation produced by a 2 V modulating signal.

Solution

(*a*) Sensitivity $= \dfrac{75 \text{ kHz}}{10 \text{ V}} = 7.5 \text{ kHz/V}$ (*Ans.*)

(*b*) $kf_d = \frac{2}{10} \times 75 \text{ kHz} = 15 \text{ kHz}$ (*Ans.*)

Alternatively,

$kf_d = 7.5 \text{ kHz/ V} \times 2 \text{ V} = 15 \text{ kHz}$ (*Ans.*)

The FREQUENCY SWING of an f.m. wave is defined by the limits between which the carrier frequency is deviated, i.e. twice the frequency deviation.

Modulation Index

The modulation index m_f of a frequency-modulated wave is the ratio of the frequency deviation of the carrier to the modulating signal frequency, i.e.

$$m_f = kf_d/f_m \qquad (2.2)$$

The modulation index is equal to the peak PHASE DEVIATION, in radians, of the carrier, and it determines the amplitudes and the frequencies of the components of the modulated wave.

Deviation Ratio

The deviation ratio D of a frequency-modulated wave is the particular case of the modulation index when *both* the frequency deviation *and* the modulating frequency are at their maximum values:

$$D = f_d/f_{m(max)} \qquad (2.3)$$

The deviation ratio is the parameter used in the design of a system and its value is fixed. Conversely, the modulation index will continually vary as the amplitude and/or frequency of the modulating signal changes.

EXAMPLE 2.3

A 100 MHz carrier wave is frequency modulated by a 10 V 10 kHz sinusoidal voltage using a linear modulator. The instantaneous carrier frequency varies between 99.95 and 100.05 MHz. Calculate (*a*) the sensitivity of the modulator, (b) the modulation index, (c) the peak phase deviation of the carrier.

Solution
(a) The peak frequency deviation is 0.05 MHz. Therefore

$$\text{Modulator sensitivity} = \frac{0.05 \times 10^6}{10} = 5.0 \text{ kHz/V} \quad (Ans.)$$

(b) From equation (2.2),

$$m_f = \frac{50 \times 10^3}{10 \times 10^3} = 5 \quad (Ans.)$$

(c) The peak phase deviation of the carrier is equal to the modulation index i.e.

5 radians   (Ans.)

EXAMPLE 2.4

What will be the new values of the peak frequency and phase deviations in the system of Example 2.3 if the amplitude and frequency of the modulating signal are changed to 20 V and 5 kHz respectively?

Solution
If the amplitude of the modulating signal voltage is doubled, the frequency deviation of the carrier will also be doubled. Therefore,

$$kf_d = 0.1 \text{ MHz} = 100 \text{ kHz} \quad (Ans.)$$

The peak phase deviation is given by kf_d/f_m. Therefore,

$$\text{Peak phase deviation} = \frac{5 \times 20/10}{5/10} = 20 \text{ radians} \quad (Ans.)$$

Frequency Spectrum of a Frequency-modulated Wave

When a sinsusoidal carrier wave of frequency f_c is frequency modulated by a sinusoidal signal of frequency f_m, the modulated wave may contain components at a number of different frequencies. These frequencies are the carrier frequency and a number of sidefrequencies positioned either side of the carrier. The sidefrequencies are spaced apart at frequency intervals equal to the modulating frequency. The first-order sidefrequencies are $f_c \pm f_m$, the second-order sidefrequencies are $f_c \pm 2f_m$, the third-order sidefrequencies are $f_c \pm 3f_m$, and so on.

The amplitudes of the various components, including the carrier, depend upon the value of the modulation index or deviation ratio, as shown by the curves given in Fig. 2.3. Only the first nine orders of sidefrequencies have been shown in order to clarify the drawing but many more are equally possible. The carrier component is zero for values of modulation index of 2.405, 5.520 and 8.654. This is in contrast with d.s.b. amplitude modulation where the carrier is always present.

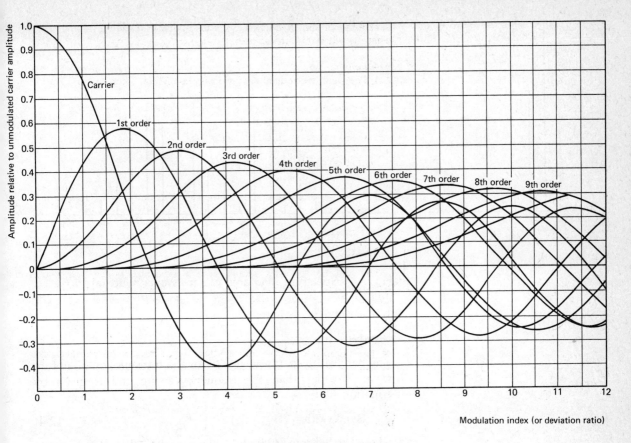

Fig. 2.3 Showing how the amplitudes of the various components of an f.m. wave vary with the modulation index

Fig. 2.3 can be used to determine the frequencies contained within a particular frequency-modulated wave. The amplitudes of each component present in the wave are obtained by projecting from the modulation index (or deviation ratio) value on the horizontal axis, onto the appropriate curve, and thence to the vertical axis. Negative signs are omitted since only the magnitude of each component is wanted.

EXAMPLE 2.5

Plot the frequency spectrum diagrams of a frequency-modulated wave having a deviation ratio of (*a*) 1 and (*b*) 5.

Solution
The required spectrum diagrams are shown in Figs. 2.4*a* and *b* respectively.
The spectrum diagrams of Fig. 2.4 show clearly that an increase in the modulation index of an f.m. wave will result in an increase in the number of sidefrequencies generated.

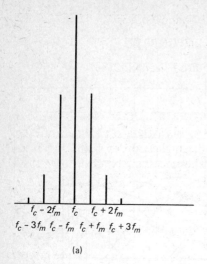

$f_c - 2f_m$ f_c $f_c + 2f_m$
$f_c - 3f_m$ $f_c - f_m$ $f_c + f_m$ $f_c + 3f_m$

(a)

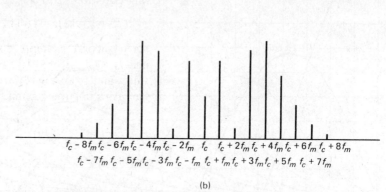

$f_c - 8f_m$ $f_c - 6f_m$ $f_c - 4f_m$ $f_c - 2f_m$ f_c $f_c + 2f_m$ $f_c + 4f_m$ $f_c + 6f_m$ $f_c + 8f_m$
$f_c - 7f_m$ $f_c - 5f_m$ $f_c - 3f_m$ $f_c - f_m$ $f_c + f_m$ $f_c + 3f_m$ $f_c + 5f_m$ $f_c + 7f_m$

(b)

Fig. 2.4 Spectrum diagrams of f.m. waves with (a) $m_f = 1$, (b) $m_f = 5$

Bandwidth

The bandwidth that is necessary for the transmission of an f.m. wave is wider than the frequency swing and is given by equation (2.4):

$$\text{Bandwidth} = 2(kf_d + f_m) \qquad (2.4)$$

where, as before, kf_d is the frequency deviation of the carrier and f_m is the modulating signal frequency. Equation (2.4) assumes that any sidefrequency whose amplitude is less than 10% of the amplitude of the *unmodulated* carrier wave need not be transmitted.

An f.m. system will, of course, be designed to transmit the most demanding modulating signal without excessive distortion. Such a signal is the one whose magnitude produces the rated system deviation and whose frequency is the maximum to be transmitted by the system. The bandwidth required for the satisfactory transmission of this signal is given by equation (2.5):

$$\text{System bandwidth} = 2(f_d + f_{m(max)}) \cdots \qquad (2.5)$$

The accuracy of equations (2.4) and (2.5) can readily be checked with the aid of Fig. 2.4. Consider, as an example, the B.B.C. v.h.f. frequency-modulated sound broadcast system; the parameters of this system include a rated system deviation of 75 kHz and a maximum modulating frequency of 15 kHz. The deviation ratio D is 5 and, from Fig. 2.4b, the highest order sidefrequency that needs to be transmitted (amplitude less than ± 0.1 on the vertical scale of Fig. 2.3) is the sixth.

This means that the necessary bandwidth is

$$f_c \pm 6f_m = 12f_m = 12 \times 15 \text{ kHz} = 180 \text{ kHz}.$$

Using equation (2.5) the required bandwidth is

$$2(75 + 15) \text{ kHz} = 180 \text{ kHz}$$

As a second example consider a narrow band system in which $f_d = f_{m(max)} = 3$ kHz. For this system the deviation ratio is unity and from Fig. 2.3a the highest-order significant side-frequencies are the second. The necessary bandwidth is

$$f_c \pm 2f_m = 4 \times 3 \text{ kHz} = 12 \text{ kHz}$$

From equation (2.5) the necessary bandwidth is

$$2(3 + 3) \text{ kHz} = 12 \text{ kHz} \quad \text{as before}$$

Power Contained in a Frequency-modulated Wave

Since the amplitude of a frequency-modulated wave does not vary, the total power contained in the wave is constant and equal to the unmodulated carrier power.

Phase Modulation

When a carrier wave is phase modulated, its instantaneous phase is made to vary in accordance with the characteristics of the modulating signal. The magnitude of the phase deviation is *proportional* to the modulating signal voltage, while the number of times per second the phase is deviated is equal to the modulating frequency.

The maximum phase deviation permitted in a phase modulation system is known as the RATED SYSTEM DEVIATION Φ_d and, as with frequency modulation, it sets an upper limit to the modulating signal voltage. A modulating voltage of lesser amplitude will produce a phase deviation equal to $k\Phi_d$, where k has the same meaning as before. The product $k\Phi_d$ is the modulation index of the phase-modulated wave. Modulating the phase of the carrier will at the same time vary the instantaneous carrier frequency. The frequency deviation produced is proportional to *both* the amplitude and the frequency of the modulating signal.

Frequency and phase modulation are sometimes grouped together and called ANGLE MODULATION since they each deviate both the frequency and the phase of the carrier voltage. The differences between the two types of modulation are tabulated in Table 2.1.

Table 2.1

Modulation	Frequency deviation	Phase deviation
Frequency	Proportional to voltage of modulating signal	Proportional to voltage and inversely proportional to frequency of modulating signal
Phase	Proportional to both voltage and frequency of modulating signal	Proportional to voltage of modulating signal

EXAMPLE 2.6

A carrier wave is angle modulated by a sinusoidal voltage and then has a phase deviation of 3 radians and a frequency deviation of 6 kHz. If the voltage of the modulating signal is doubled and the modulating frequency is reduced by half, the frequency deviation is unaltered. Is this frequency- or phase-modulation? What is the new phase deviation?

Solution
Refering to Table 2.1 it is clear that the carrier has been *phase modulated.* (*Ans.*)

New phase deviation $= 3 \times 2 = 6$ rad (*Ans.*)

The frequency spectrum of a phase-modulated wave is exactly the same as that of the frequency-modulated wave having the same numerical value of modulation index.

Signal-to-Noise Ratio in F.M. Systems

During its transmission, a frequency modulated signal will be subjected to noise and interference voltages. The effect of these unwanted voltages is to vary both the amplitude and the phase of the f.m. signal. The amplitude variations thus produced have no effect on the performance of the system since they will have been removed by a *limiter* circuit in the radio receiver. The phase deviation of the signal, however, means that the carrier is effectively frequency modulated by the noise, and a noise voltage will appear at the output of the radio receiver.

The magnitude of the output noise voltage is directly proportional to frequency and gives rise to the TRIANGULAR NOISE SPECTRUM (Fig. 2.5). The output noise voltage rises linearly from zero frequency until, theoretically, at a frequency equal to the rated system deviation f_d, it is equal to the noise output voltage from an a.m. system subject to the same noise/interfering voltage. For many systems not all of this noise is able to pass through the audio stage of the receiver, since the frequency deviation is larger than the audio passband.

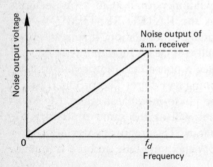

Fig. 2.5 Triangular noise spectrum of an f.m. system

As an example, consider the v.h.f. sound broadcast system of the B.B.C.; this system employs a rated system deviation of 75 kHz and a maximum modulating frequency of 15 kHz, i.e. the deviation ratio D is 5. Then, refering to Fig. 2.6, the areas, enclosed by the points ABC and ABDE represent, respectively, the output noise voltages of the f.m. and the a.m. systems. Clearly, the noise output of the f.m. receiver is smaller than that of the a.m. receiver. This means that frequency modulation can provide an increase in the output signal-to-noise ratio of a system. This is one of the advantages of frequency modulation over d.s.b. amplitude modulation.

The size of the signal-to-noise ratio improvement depends upon the rated system deviation used and hence upon the system bandwidth available. If the frequency deviation of the B.B.C. broadcast system were reduced to 15 kHz without changing the maximum modulating frequency, so that $D = 1$, the output noise-voltage would be represented by the area enclosed by the points A,B,D in Fig. 2.6. Area ABD is larger than area ABC, an indication that the reduction in frequency deviation has resulted in an increase in the output noise voltage. The signal-to-noise ratio† improvement of an f.m. system over an a.m. system is given by equation (2.6):

$$\text{Signal-to-noise ratio increase} = 20 \log_{10} D\sqrt{3} \text{ dB} \qquad (2.6)$$

where D is the deviation ratio.

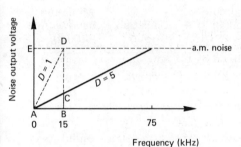

Fig. 2.6 Illustrating the effect on the noise output of an f.m. system of reducing the deviation ratio

EXAMPLE 2.7

A frequency-modulation system has an output signal-to-noise ratio of 30 dB when the deviation ratio is 3.5. What will be the output signal-to-noise ratio if the deviation ratio is increased to 5?

Solution

$$\text{New output signal-to-noise ratio} = 30 + 20 \log_{10}(5/3.5)$$
$$= 33.1 \text{ dB} \qquad (Ans.)$$

The signal-to-noise ratio at the output of a frequency-modulation system is a function of the rated system deviation chosen. An increase in the deviation will increase the output signal-to-noise-ratio but, at the same time (see equation (2.5)), also increase the required system bandwidth. Thus, the choice of rated system deviation for a particular system must be a compromise between the conflicting requirements of maximum output signal-to-noise ratio and minimum bandwidth.

For its v.h.f. sound broadcasts, the B.B.C. use, as previously mentioned, a rated system deviation of 75 kHz which gives a

† Signal-to-noise ratio = (Wanted signal power)/(Unwanted noise power) [see EIII]

deviation ratio of 5 and a minimum bandwidth requirement of 180 kHz. The sound section of u.h.f. television transmissions use a rated system deviation of 50 kHz. This gives a deviation ratio of 3.33 and a necessary bandwidth of 130 kHz. The rated system deviation chosen for mobile systems is always considerably smaller than the B.B.C. figures. This is because the need for minimum channel bandwidth is of paramount importance, while a wide audio-frequency response is not necessary.

Some typical figures are given in Table 2.2.

Table 2.2

Rated system deviation (kHz)	Maximum modulating frequency (kHz)	Deviation ratio	Minimum bandwidth (kHz)
2.5	3.5	0.71	12
5	3.4	1.47	16.8
2.8	3.4	0.82	12.4

Multi-channel telephony systems are often routed over frequency-modulated *radio-relay systems* operating in the s.h.f. band. The frequency deviation of such systems is usually quoted as a number of megahertz peak-to-peak deviation. The systems are generally set up to give the frequency deviations listed in Table 2.3.

Table 2.3

System	Frequency deviation	Note
600 channel	200 kHz	Test tone on one channel
1800 channel	140 kHz	"
Carrying t.v. signal	8 MHz	1 V p-p Input

Pre-emphasis and De-emphasis

Most waveforms transmitted by communication systems contain a large number of components at different frequencies. Usually the higher-frequency components are of smaller amplitude than the components at lower frequencies. For example, the frequencies contained in a speech waveform mainly occupy the band 100–10 000 Hz but most of the power is contained at frequencies in the region of 500 Hz for men and 800 Hz for women. Since the noise appearing at the output of a frequency-modulated system increases linearly with increase in frequency, the signal-to-noise ratio falls at high frequencies.

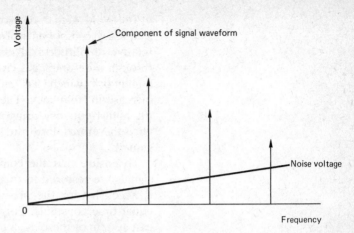

Fig. 2.7 Signal-to-noise ratio at the output of an f.m. system

This is shown by Fig. 2.7 in which a signal containing components at five different frequencies has been assumed. For a multi-channel system this means that the signal-to-noise ratio will be worse in the highest-frequency channel. To improve the signal-to-noise ratio at the higher frequencies, pre-emphasis of the modulating signal is applied at the transmitter.

Refer to Fig. 2.8. The modulating signal is passed through a PRE-EMPHASIS network which accentuates the amplitudes of the high-frequency components of the signal relative to the low-frequency components, before it is applied to the radio transmitter. During its transmission from transmitter to receiver, noise and interference will be superimposed upon the signal so that the output of the radio receiver will exhibit the triangular noise spectrum. Now, however, the signal-to-noise ratio at the higher frequencies is greater than it would have been without pre-emphasis.

Fig. 2.8 The effect of pre-emphasis on the output signal-to-noise ratio of an f.m. system

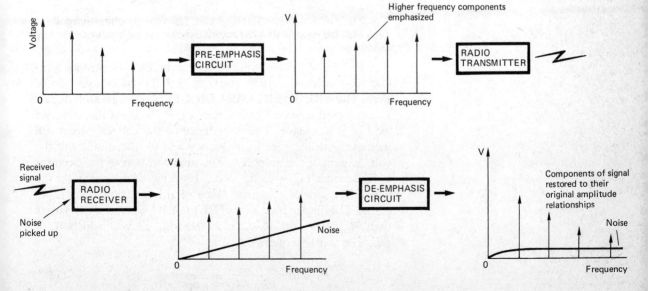

To avoid signal distortion it is necessary to restore the frequency components of the received signal to their original relative amplitude relationships. For this the signal is passed through a de-emphasis circuit. The DE-EMPHASIS circuit is a network which has an attenuation which increases with increase in frequency. The de-emphasis circuit also attenuates the high-frequency components of the noise voltage and does not, therefore, lose the signal-to-noise ratio improvement gained.

To ensure that the component frequencies of the received signals are restored to their original amplitude relationships, it is necessary for the pre- and de-emphasis networks to have equal time constants. For example, in the v.h.f. sound broadcast system of the U.K. a time constant of 50 μs is employed.

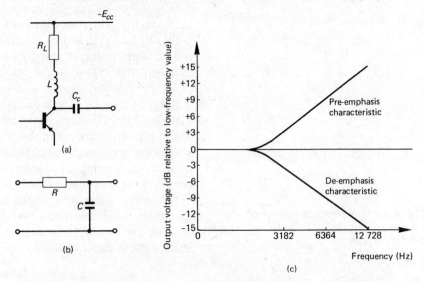

Fig. 2.9 (a) A pre-emphasis circuit, (b) a de-emphasis circuit, (c) pre- and de-emphasis characteristics for 50 μs sound broadcast systems

A variety of different networks can be used to provide pre- and de-emphasis circuits and Fig. 2.9 shows an example of each. The PRE-EMPHASIS CIRCUIT consists of an inductor L connected in series with a resistor R_L to form the collector load of a transistor. The impedance of the collector load will increase with increase in frequency and so, therefore, will the voltage gain of the amplifier. The time constant of the circuit is L/R_L seconds (C_c is merely a d.c. blocking component). At the lower frequencies the impedance of the collector load is R_L ohms. The impedance $Z = \sqrt{[R_L^2 + (\omega L)^2]}$ will be 3 dB larger than R_L ohms, i.e. $R_L\sqrt{2}\Omega$, at the frequency at which $R_L = \omega L$. The 3 dB frequency f_{3dB} is

$$f_{3dB} = R_L/2\pi L = \frac{1}{2\pi \times (\text{time constant})} \tag{2.7}$$

Therefore,

$$f_{3dB} = \frac{1}{2\pi \times 50 \times 10^{-6}} = 3182 \text{ Hz}$$

At frequencies higher than f_{3dB} the impedance of the collector load and hence the output voltage will double for each twofold increase in frequency, i.e. increase at a rate of 6 dB/octave. This means that the impedance is 9 dB greater at 6364 Hz and 15 dB greater at 12 728 Hz (see Fig. 2.9c). The output-voltage/frequency characteristic of the de-emphasis circuit must be the inverse of this and is also shown in the figure.

The improvement in output signal-to-noise ratio produced by the use of pre-emphasis is not easy to assess but for sound broadcasting is generally supposed to be about 6 dB.

A mobile communication radio system will have an audio passband of about 300–3000 Hz and a typical pre-emphasis characteristic is 6 dB per octave over this band.

The choice of pre-emphasis characteristic to be used for a multi-channel radio-relay system has been made by the C.C.I.R. and is given in Fig. 2.10a. Signals carried by the low-frequency channels are reduced in amplitude, while signals applied to high-frequency channels have their amplitude increased; in the case of the top channel by 4 dB. The standard frequency deviation is produced at a baseband frequency of 60.8% of the maximum.

Fig. 2.10 (*a*) Pre-emphasis characteristics for multi-channel radio-relay systems

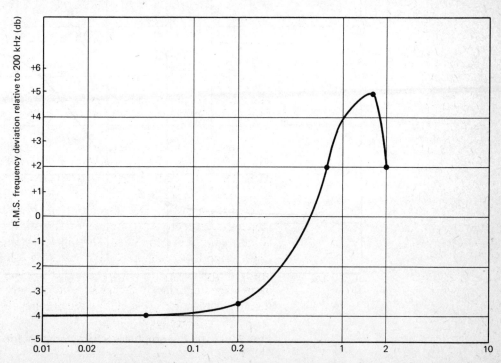

R.M.S. frequency deviation relative to 200 kHz (db)

Normalized frequency (frequency/highest frequency in baseband)

Many wideband s.h.f. radio systems carry television signals instead of multi-channel telephony. The television signal is also pre-emphasized but the purpose of the operation is now to make it possible for the same modulator to be used for both telephony and television. The television signal waveform is of non-symmetrical shape, since most of its energy is contained at low frequencies and the pre-emphasis network reduces the amplitudes of the low-frequency components. The t.v. pre-emphasis characteristic is shown in Fig. 2.10*b*; standard frequency deviation is produced at 1.6 MHz.

Relative Merits of Amplitude, Frequency and Phase Modulation

The advantages of frequency modulation over d.s.b. amplitude modulation are listed below:

(*a*) The dynamic range (the range of modulating signal amplitudes from lowest to highest) provided is much greater.

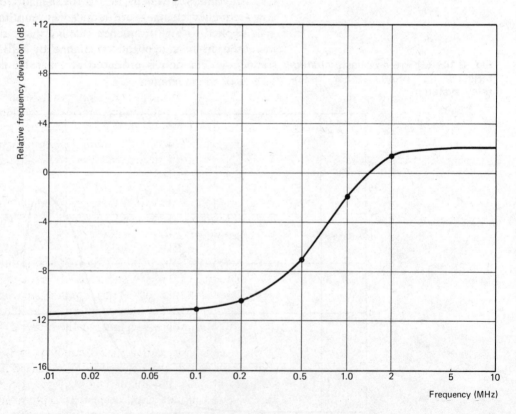

Fig. 2.10 (*b*) Pre-emphasis characteristics for u.h.f. television signals

(b) The frequency-modulation transmitter is more efficient. There are two reasons for this; firstly, Class C amplifiers can be used throughout the r.f. section of the transmitter, and secondly, since the amplitude of an f.m. wave is constant each r.f. stage in the transmitter can be operated in its optimum manner. This is not the case in a d.s.b. a.m. transmitter because each of the stages must be capable of handling a *peak* power which can be considerably larger than the *average* power.

(c) Since an f.m. receiver does not respond to any amplitude variations of the input signal, selective fading is not a problem.

(d) The use of frequency modulation provides an increase in the output signal-to-noise ratio of the radio receiver provided a deviation ratio greater than unity is used. Narrowband f.m. systems do not share this advantage.

(e) An f.m. receiver has the ability to suppress the weaker of two signals which are simultaneously present at its aerial terminals at or near the same frequency. The CAPTURE RATIO is expressed in dB; the lower the value the better. For example, a capture ratio of 4.5 dB means that, if the receiver is tuned to a particular signal, it will not respond to any other signal whose amplitude is 4.5 dB or more below the amplitude of the wanted signal.

(f) When a multi-channel telephony system is transmitted over a radio-relay link, linearity in the output/input transfer characteristic of the equipment is of the utmost importance in order to minimize inter-channel crosstalk. The required linearity is easier to obtain using frequency modulation.

Although phase modulation is very similar to frequency modulation, it is rarely used for analogue systems. There are two reasons for this;

(a) Frequency modulation is more efficient than phase modulation in its use of the available frequency spectrum.

(b) Demodulation of a phase-modulated wave is more difficult than f.m. demodulation since it requires a very stable reference oscillator in the receiver.

The main disadvantage of frequency modulation is, of course, the much wider bandwidth required if the possible signal-to-noise ratio improvement is to be realized. For narrow-band mobile applications the capture effect may also prove to be disadvantageous since when a mobile receiver is near the edge of the service area it may be captured by an unwanted signal or a noise voltage.

Measurement of a Frequency-modulated Wave

The parameter of a frequency-modulated wave that is usually measured is the frequency deviation. Commercial f.m. deviation meters are available but the measurement can be carried out by the CARRIER DISAPPEARANCE method. The amplitude of the carrier frequency component of an f.m. wave is a function of the modulation index. The carrier voltage is zero for values of modulation index of 2.405, 5.52, 8.65, etc. If, for any one of these modulation indexes, the modulation frequency is known, the frequency deviation can be calculated.

To measure the frequency deviation of an f.m. wave the signal is applied to an instrument known as a *spectrum analyzer*. The spectrum analyzer is an instrument which displays voltage to a base of frequency (as opposed to a c.r.o. which displays voltage to a base of time). The spectrum analyzer therefore displays the spectrum diagram of the f.m. signal (see Fig. 2.4 for example). It is adjusted to display only the carrier and the first-order sidefrequencies.

With the modulating frequency kept at a constant value, the amplitude of the modulating signal is increased from zero, which varies the frequency deviation until the carrier first goes to zero. Then $m_f = 2.405$ and the frequency deviation can be calculated. Further increase in the modulating signal voltage will cause the carrier component to reappear and then again go to zero when the modulation index becomes 5.52.

EXAMPLE 2.8

In a measurement of the frequency deviation of an f.m. signal, the frequency of a signal generator was set to 3 kHz. Calculate the frequency deviation at (*a*) the first and (*b*) the second carrier disappearance.

Solution

(*a*) $m_f = 2.405 = \dfrac{kf_d}{3 \times 10^3}$

$kf_d = 2.405 \times 3 \times 10^3 = 7.215$ kHz   (*Ans.*)

(*b*) $kf_d = 5.52 \times 3 \times 10^3 = 16.56$ kHz   (*Ans.*)

Exercises

2.1. (*a*) What is meant by the following terms in connection with frequency modulation: (i) modulation index, (ii) frequency deviation, (iii) practical bandwidth? (*b*) When the modulation index of a certain f.m. transmitter is 7 in a practical bandwidth of 160 kHz, what is its frequency deviation?

2.2. (*a*) What is the meaning of the term modulation index when applied to an f.m. signal? (*b*) What is the meaning of the term modulation factor when applied to an a.m. signal? (c) Describe a method of measuring each.   (C & G)

2.3. (a) What do you understand by the following terms: (i) frequency deviation, (ii) modulation index, (iii) deviation ratio? (b) The r.f. bandwidth required for an f.m. transmitter is 100 kHz when the modulation index is four. If the modulation signal level is increased by 6 dB, what is (i) the new modulation index, (ii) the bandwidth required?   (C & G)

2.4. (a) Briefly explain the purpose of pre-emphasis and de-emphasis in f.m. systems. (b) What is the approximate improvement in output signal-to-noise ratio when pre-emphasis and de-emphasis circuits are used? (c) Draw the circuit diagram of (i) a pre-emphasis and (ii) a de-emphasis circuit. (d) Indicate the components in these circuits upon which the degree of emphasis depends.   (C & G)

2.5. (a) Why is the use of frequency modulation confined to the v.h.f. band and above as a general rule? (b) Briefly describe what you understand by the term capture effect. (c) (i) Indicate typical values of frequency deviation and highest modulating frequency used in f.m. broadcasting services operating in the v.h.f. band; (ii) Using these values calculate the maximum percentage band occupancy when the minimum carrier spacing is 2.2 MHz for transmitters serving the same area.   (C & G)

2.6. When a certain sinusoid is used to frequency modulate a v.h.f. carrier, the required bandwidth is 200 kHz. It is desired to retain the same modulation index while reducing the necessary bandwidth to 100 kHz. What changes should be made to the input to the modulator to achieve the required deviation?
                                                          (part C & G)

2.7. Describe, using sketches where necessary, how the amplitude and frequency of a modulating signal are conveyed by (a) amplitude modulation, (b) frequency modulation. Discuss briefly the advantages and disadvantages of f.m. compared with a.m. in a v.h.f. communication system. The r.f. bandwidth of an f.m. transmitter is 80 kHz when a 6 kHz modulating signal is applied. What bandwidth is required if the modulating signal level is reduced by 6 dB?

2.8. Explain why f.m. transmission can give an improved signal-to-noise ratio compared with a.m. transmission of the same carrier power. What characteristics of f.m. transmission determine the magnitude of this improvement? An f.m. radio link having a deviation ratio of 10 is to transmit speech occupying the audio band up to 3 kHz. What r.f. bandwidth would normally be used for this transmission? What would be the effect on (a) the r.f. bandwidth and (b) the signal-to-noise ratio if the deviation ratio were reduced to 5?   (C & G)

2.9. What is the meaning of the term dynamic range when used in conjunction with a modulation system? Explain why it is possible for an f.m. system to have a greater range than an a.m. system.
                                                          (C & G)

2.10. Use the graph given in Fig. 2.3 to draw the spectrum diagram of a 80 MHz carrier wave frequency modulated with an index of 4. What bandwidth is required?

Short Exercises

2.11. Draw the circuit diagram of (*a*) a pre-emphasis circuit and (*b*) a de-emphasis circuit.

2.12. Why will the full advantages of frequency modulation not be realised unless the signal at the output of the i.f. amplifier is large?

2.13. Why must frequency modulation be used in conjunction with a very high frequency carrier?

2.14. With which of the following modulation techniques is the triangular spectrum of noise associated: (i) a.m., (ii) f.m., (iii) both a.m. and f.m. (*b*) At which of the following does the triangular spectrum of noise appear: (i) r.f., (ii) i.f., (iii) a.f.?

2.15. The r.f. bandwidth required by an f.m. transmitter is 120 kHz when the modulation index is 3. What bandwidth is needed if the modulation index is increased six times?

2.16. Briefly explain the differences between frequency modulation and phase modulation.

2.17. Define the following terms used with frequency modulation: (i) frequency deviation, (ii) rated system deviation, (iii) modulation index, and (iv) deviation ratio.

2.18. What characteristic of a frequency-modulated wave determines (i) the amplitude and (ii) the frequency of the audio output of an f.m. receiver?

2.19. An f.m. transmitter has a frequency swing of 80 kHz. Determine its frequency deviation when the modulating signal voltage is halved.

2.20. Explain why frequency modulation is not employed for medium-wave sound broadcasting.

3 Modulators and Demodulators

Introduction

Any radio system must operate in the frequency band that has been allocated to it. This means that the modulating, or baseband, signal must be frequency translated to a different part of the frequency spectrum. The translation process is carried out in the radio transmitter by modulating, either in amplitude or in frequency, a carrier wave of appropriate frequency. In the radio receiver the reverse process must be carried out, i.e. the signal must be demodulated. In this chapter the operation of the more commonly used modulators, excepting those using Class C power amplifiers, and demodulators will be considered.

Amplitude Modulators

The modulator circuits used in d.s.b radio transmitters must permit the modulating signal to amplitude modulate the carrier wave without the production of an excessive number of extra, unwanted frequencies. The modulator used in an s.s.b. system must, in addition, also suppress the carrier frequency and, in some cases, the modulating signal also.

D.S.B. Modulators

The majority of d.s.b. amplitude-modulated radio transmitters use an anode- or collector-modulated Class C r.f. tuned power amplifier circuit and these will be discussed in Chapter 7.

Other d.s.b. modulators utilize the non-linear relationship between applied voltage and resulting current of many electronic devices. If a carrier wave at frequency f_c and a sinusoidal modulating signal at frequency f_m are applied in series to a non-linear device, the resultant current will contain components at various frequencies. Amongst these are components at the carrier frequency f_c and the sum and difference frequencies $f_c \pm f_m$. If these components are selected, and all others are rejected, by means of a parallel-tuned circuit, an amplitude-modulated wave will be obtained. The non-linear device can be a diode but is more likely to be a suitably biased bipolar or field-effect transistor.

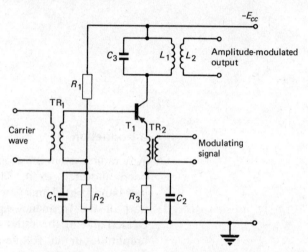

Fig. 3.1 D.S.B. amplitude modulator

One possible transistor modulator circuit is shown in Fig. 3.1. The transistor is biased to operate over the non-linear part of its mutual characteristics. The carrier and modulating signal voltages are introduced into the base/emitter circuit of T_1 by means of transformers TR_1 and TR_2 respectively. The collector current contains the wanted carrier and sidefrequency components plus various other, unwanted components. The collector circuit is tuned to the carrier frequency and has a selectivity characteristic such that the required amplitude modulated waveform appears across it. The various unwanted components are at frequencies well removed from resonance and do not develop a voltage across the collector load. The use of a non-linear modulator is restricted to low-power applications because the method has the disadvantages of low efficiency and a high percentage distortion level.

S.S.B. Modulators

In an s.s.b. or an i.s.b. system the carrier component is suppressed during the modulation process by using a *balanced modulator*. When a low-level pilot carrier is transmitted, it is added to the s.s.b.s.c. signal at a later point in the transmitter.

The circuit of a transistor balanced modulator is given in Fig. 3.2. Transistors T_1 and T_2 are biased to operate on the non-linear part of their characteristics. Since the input transformer TR_1 is centre-tapped, the modulating signal voltages applied to transistors T_1 and T_2 are in antiphase with one another. The carrier voltage is introduced into the circuit between the centre-tap on the input transformer and earth, and so applies in-phase voltages to the two transistors. The collector currents of each transistor contain components at a number of different frequencies, and flow in opposite direc-

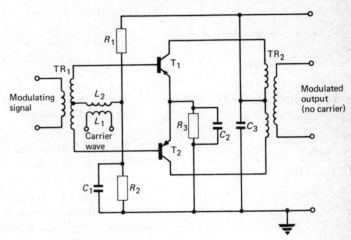

Fig. 3.2 Transistor balanced modulator

tions in the primary winding of the output transformer TR_2. The phase relationships of the various components of the collector currents are such that the current flowing in the secondary winding of TR_2 contains components at the modulating frequency and at the upper and lower sidefrequencies but *not* at the carrier frequency. In practice, the two halves of the circuit do not have identical characteristics and some *carrier leak* is always present at the output of the circuit.

Many balanced modulators, particularly those employed in multichannel line systems, do not utilize the square-law characteristics of a diode or transistor but instead use the device as an electronic switch. When a diode or transistor is forward biased its resistance is low, and when it is reverse biased its resistance is high. Provided the carrier voltage is considerably greater than the modulating signal voltage, the carrier will control the switching of the device. Ideally, a device should have zero forward resistance and infinite reverse impedance, and this will be assumed in the circuits that follow.

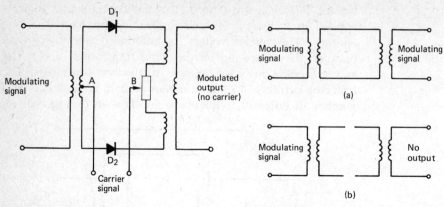

Fig. 3.3 Single balanced modulator

Fig. 3.4 Operation of a single balanced modulator

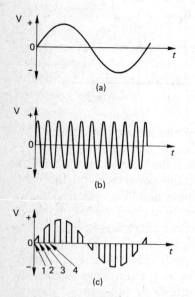

Fig. 3.5 Output waveform of a single balanced modulator: (a) modulating signal, (b) carrier wave, (c) output waveform

(1) Fig. 3.3 shows the circuit of a SINGLE-BALANCED DIODE MODULATOR. During the half-cycles of the carrier waveform that make point A positive with respect to point B, diodes D_1 and D_2 are forward biased and have zero resistance. The modulator may then be redrawn as shown in Fig. 3.4a; obviously the modulating signal will appear at the output terminals of the circuit. Similarly, when point B is taken positive relative to point A, the diodes are reverse biased and Fig. 3.4b represents the modulator. The action of the modulator is to switch the modulating signal on and off at the output terminals of the circuit. The output waveform of the modulator can be deduced by considering the modulating signal and carrier waveforms at different instants. Consider Fig. 3.5: during the first positive half-cycle of the carrier wave a part of the modulating signal appears at the output (1–2 in Fig. 3.5c); in the following negative half-cycle the modulating signal is cut off (2–3); in the next positive half-cycle the corresponding part of the modulating signal again appears at the output terminals (3–4); and so on.

Analysis of the output waveform shows that it contains the upper and lower sidefrequencies of the carrier ($f_c \pm f_m$), the modulating signal f_m, and a number of higher, unwanted frequencies, but the carrier component is *not* present. In practice, of course, non-ideal diodes are employed and this has the effect of generating further unwanted frequencies and of reducing the amplitude of the wanted sidefrequency. Some carrier leak also occurs, and a potentiometer is often included to enable adjustment for minimum leak to be carried out.

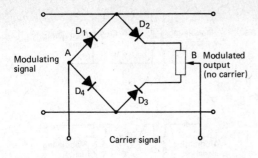

Fig. 3.6 Cowan modulator

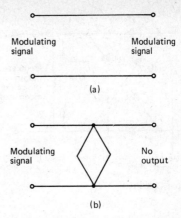

(a)

(b)

Fig. 3.7 Operation of the Cowan modulator

(2) Anotner circuit that performs the same function is the COWAN MODULATOR (Fig. 3.6). The carrier voltage is applied across points A and B and switches the four diodes rapidly between their conducting and non-conducting states. When point B is positive with respect to point A, all four diodes are reverse biased and the modulator may be represented by Fig. 3.7a; during the alternate carrier half-cycles Fig. 3.7b applies. The modulator output therefore consists of the modulating signal switched on and off at the carrier frequency. The output waveform is the same as that of the previous circuit (Fig. 3.5) and contains the same frequency components. The Cowan modulator, however, does not require centre-tapped transformers and it is therefore cheaper. It also possesses a self-limiting characteristic (i.e. the sidefrequency output voltage is proportional to the input signal level only up to a certain value and thereafter remains more or less constant).

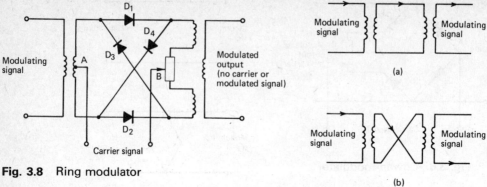

Fig. 3.8 Ring modulator

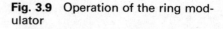

(a)

(b)

Fig. 3.9 Operation of the ring modulator

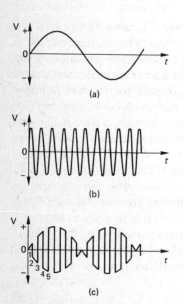

Fig. 3.10 Output waveform of the ring modulator: (a) modulating signal, (b) carrier wave, (c) output waveform

(3) Sometimes it is necessary to suppress the modulating signal as well as the carrier wave during the modulation process, and then a DOUBLE BALANCED MODULATOR is used. Fig. 3.8 shows the circuit of a *ring modulator*. During half-cycles of the carrier wave when point A is positive relative to point B, diodes D_1 and D_2 are conducting and diodes D_3 and D_4 are not; D_1 and D_2 therefore have zero resistance and D_3 and D_4 have infinite resistance; Fig. 3.9a applies. Whenever point B is positive with respect to point A, diodes D_1 and D_2 are non-conducting, D_3 and D_4 are conducting, and Fig. 3.9b represents the modulator. It is evident that the direction of the modulating signal current at the modulator output terminals is continually reversed at the carrier frequency.

The output waveform of a ring modulator is shown in Fig. 3.10c and can be deduced from Figs. 3.10a, b. Whenever the carrier voltage is positive, the modulating signal appears at the modulator output with the same polarity as *a* (see points 1–2 and 3–4 at *c*). Whenever the carrier voltage is negative the polarity of the modulating signal is reversed (points 2–3 and 4–5).

Analysis of the output waveform shows the presence of components at the upper and lower sidefrequencies of the carrier wave and a number of higher, unwanted frequencies. *Both* the carrier and the modulating signal are suppressed.

(4) The balanced modulator is also available in integrated circuit form. Besides the usual advantages of integrated circuits over discrete circuitry, the i.c. modulators also offer exceptionally good carrier suppression, fully balanced input and output circuits, and are capable of operation over a wide

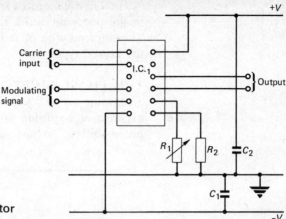

Fig. 3.11 Integrated balanced modulator

frequency band, typically up to about 100 MHz. Fig. 3.11 shows the basic circuit of an INTEGRATED DOUBLE-BALANCED MODULATOR. The variable resistor R_1 is provided for adjustment of the carrier leak appearing at the output terminals to a minimum value. R_2 is a bias component and capacitors C_1 and C_2 decouple the positive and negative power supply lines.

When a complex modulating signal is applied to any of the modulator circuits described, upper and lower sidebands are produced. In an s.s.b.s.c. system only one of the sidebands is to be transmitted and the other is to be suppressed. Two methods of sideband suppression are available, known respectively as the *filter method* and the *phasing method*.

Sideband Suppression

THE FILTER METHOD   The more commonly used method of removing the unwanted sideband is to pass the output of the balanced modulator through a band-pass filter as shown in Fig. 3.12. When the modulating signal frequency is close to the wanted sideband, the filter may not be able to provide sufficient rejection. When this is the case a double-balanced modulator is used so that the modulating signal will not appear at the input terminals of the filter.

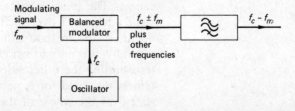

Fig. 3.12 Filter method of producing an s.s.b.s.c. signal

The filter is required to have a flat attenuation characteristic in the passband and a rapidly increasing attenuation outside. The filter may be of inductor/capacitor construction or be of the crystal type.

THE PHASE SHIFT METHOD The phasing method of generating an s.s.b.s.c. signal avoids the use of a filter at the expense of requiring an extra balanced modulator and two phase-shifting circuits. The block schematic diagram of the

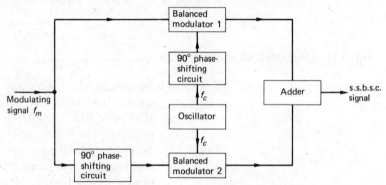

Fig. 3.13 Phasing method of producing an s.s.b.s.c. signal

phasing method is given in Fig. 3.13. Balanced modulator 1 has the modulating signal at frequency f_m applied to it, together with the output of the carrier-generating oscillator after it has been passed through a network which introduces a 90° phase lead. The other balanced modulator (2) has the carrier voltage applied directly to it but the modulating signal is first passed through a circuit which advances all frequencies by 90°. The result of the phase shifts given to the carrier and modulating signals in different parts of the circuit is that the upper sideband outputs of the two modulators are in phase, but the lower sideband outputs are in antiphase. The modulator outputs are combined in an additive circuit with the result that the lower sidebands cancel to leave a single sideband suppressed-carrier signal.

If the lower sideband is to be transmitted instead of the upper sideband, the phasing of the circuit must be altered so that the phase-shifted signal and carrier voltages are *both* applied to the same modulator. Then the upper sidebands cancel. The phasing system of s.s.b.s.c. a.m. generation has some advantages over the filter method. Firstly, since a filter is not used, the method will operate at higher frequencies and, secondly, it is easy to switch from transmitting one sideband to transmitting the other sideband. The disadvantage of the system is the need for a network that can introduce 90° phase shift over the whole of the audio-frequency band.

Frequency Modulators

The function of a frequency modulator is to vary the frequency of the carrier in accordance with the characteristics of the modulating signal applied to it. There are two fundamentally different approaches to the problem. The frequency of an inductor-capacitor oscillator can be modulated by varying the capacitance or the inductance of its frequency-determining tuned circuit. Alternatively, a crystal oscillator can be phase-modulated in such a way that a frequency-modulated wave is produced.

Direct Frequency Modulation

The basic principle upon which the operation of all direct frequency modulators is based is shown in Fig. 3.14. A circuit whose reactance, generally capacitive, can be controlled by the modulating signal is connected in parallel with the frequency-determining tuned-circuit L_1-C_1 of the oscillator. When the modulating signal voltage is zero, the effective capacitance C_e of the variable-reactance circuit is such that the oscillation frequency is equal to the nominal (unmodulated) carrier frequency, i.e.

$$f_{osc} \simeq 1/2\pi\sqrt{[L_1(C_1 + C_e)]}\,\text{Hz}$$

When the modulating signal is applied, the effective capacitance of the reactance circuit will be varied and this, in turn, will frequency-modulate the oscillator. Most direct frequency modulators are either some form of *reactance frequency modulator* or *a varactor diode modulator*.

(1) The circuit of a TRANSISTOR REACTANCE MODULATOR is shown in Fig. 3.15. Resistors R_1, R_2, R_3 and capacitor C_1 are the usual bias and decoupling components. Capacitor C_3 is a d.c. blocking component which is necessary to prevent L_2 shorting the d.c. collector potential of T_1 to earth. Inductor L_1 is a radio-frequency choke. The output

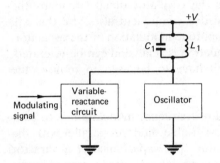

Fig. 3.14 The principle of a frequency modulator

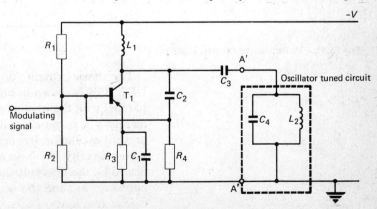

Fig. 3.15 Transistor reactance modulator

impedance of the circuit is the ratio voltage/current at the terminals AA′, and for this to be a capacitive reactance the current must lead the voltage by 90°. This is the purpose of the components R_4 and C_2; their values are chosen so that the voltage developed across R_4 makes T_1 conduct an alternating current that leads the collector voltage by very nearly the required 90°.

Frequency modulation of the oscillator frequency requires that the effective capacitance of the circuit is varied by the modulating signal. Since the capacitance is directly proportional to the mutual conductance of the transistor, it can be varied by applying the modulating signal to the base of T_1. The impedance shunted across the oscillator tuned circuit by the modulator will have a resistive component also, and this will lead to some unwanted amplitude modulation of the oscillator. Often this amplitude modulation is small and can be tolerated. If it cannot, a limiter will have to be used to remove the amplitude variations.

(2) An alternative method of frequency modulation is to connect a varactor, or voltage-variable diode in parallel with the tuned circuit of the oscillator. The capacitance of a varactor diode is a function of the reverse-biased voltage applied across it and can therefore be varied by the modulating signal. It is necessary for the oscillator frequency to be varied either side of its unmodulated value and so the varactor diode must have a mean capacitance value established by a bias voltage.

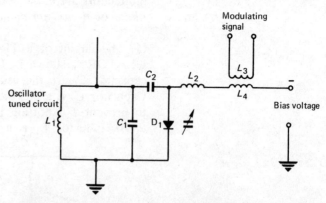

Fig. 3.16 Varactor diode modulator

The basic circuit of a VARACTOR DIODE MODULATOR is shown in Fig. 3.16. The diode D_1 is connected in parallel with the oscillator tuned circuit C_1-L_1. Capacitor C_2 is merely a d.c. block and L_2 is a radio-frequency choke to prevent oscillation frequency currents reaching the modulating signal circuitry. With no modulating signal the diode is reverse biased by the bias voltage V_B and provides the capacitance C_d necessary to tune the oscillator to the required unmodulated carrier frequency, i.e.

$$f_{osc} = 1/2\pi\sqrt{[L(C_1 + C_d)]}\,\text{Hz} \qquad (3.1)$$

When the modulating signal is applied, a voltage $V_m \sin \omega_m t$ appears across inductor L_4 and the total voltage applied to the diode is

$$-V_B + V_m \sin \omega_m t \text{ volts}$$

This voltage varies the diode capacitance and in so doing frequency modulates the oscillator.

EXAMPLE 3.1

(*a*) A variable capacitance diode has a characteristic given by Table 3.1. Plot a graph of diode capacitance against diode voltage.

Table 3.1

Reverse voltage (V)	−1	−2	−3	−4	−5	−6
Diode capacitance (pF)	12.5	7.5	6.0	5.0	4.3	3.8

(*b*) A 90 MHz oscillator employs a parallel-tuned circuit to control the frequency of oscillation. The circuit consists of a coil of inductance $0.2\,\mu\text{H}$ in parallel with a 10 pF capacitor, across which is connected the variable-capacitance diode described in (*a*) above. Using the given characteristic, determine the voltage which must be applied to the diode for oscillations to occur at 90 MHz

(*C & G*)

Solution
(*a*) The required graph is given in Fig. 3.17.

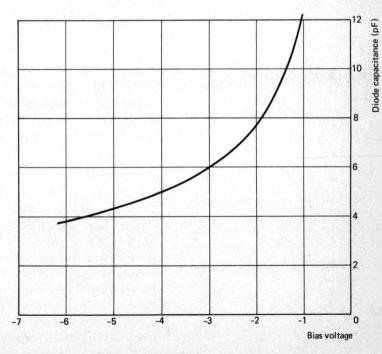

Fig. 3.17

Bias voltage

(*b*) The total capacitance needed to tune the inductance to resonance at 90 MHz is

$$C_t = 1/(4\pi^2 \times 90^2 \times 10^{12} \times 0.2 \times 10^{-6}) = 15.63 \, \text{pF}$$

Therefore, the diode must provide a capacitance of 5.63 pF. From the graph of Fig. 3.17 this capacitance value is obtained when the applied voltage is -3.3 V. (*Ans.*)

Indirect Frequency Modulation

Direct frequency modulation of an oscillator has one main disadvantage. Since an inductor/capacitor oscillator must be used, the inherent frequency stability of the unmodulated carrier frequency is not high enough to meet modern requirements. There are two possible solutions to this problem. Firstly a direct modulation system can be used and *automatic frequency control* applied to the transmitter or, secondly, a crystal oscillator can be used. With the second solution, an indirect method of modulation must be used since the frequency of a crystal oscillator cannot be varied.

The phase deviation of a frequency-modulated signal is proportional to the amplitude of the modulating signal and inversely proportional to the modulating frequency. The phase deviation of a phase-modulated signal is proportional to the modulating signal voltage only. The relationships mean that if the modulating signal is *integrated* and is then used to *phase modulate* a carrier wave a *frequency-modulated* waveform is

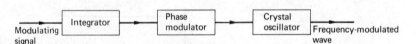

Fig. 3.18 Use of a phase modulator to generate a frequency-modulated wave

obtained. The arrangement is shown in block schematic form in Fig. 3.18.

One type of phase modulator that has enjoyed considerable popularity is the ARMSTRONG CIRCUIT shown in Fig. 3.19. The output voltage of the balanced modulator contains

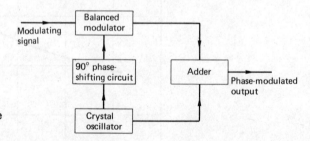

Fig. 3.19 The Armstrong phase modulator

the upper and lower sidebands produced by amplitude modulating the 90° phase-shifted carrier with the modulating signal. The s.s.b.s.c. amplitude-modulated signal is then added to the zero-phase-shifted carrier component, to produce a phase-modulated signal. If the modulating signal is integrated before it arrives at the balanced modulator, a frequency-modulated output is produced. The frequency deviation produced is very small, being usually of the order of about 30 Hz.

Many alternative phase modulator circuits have been invented but mention will only be made of one of them. Phase-modulation of a carrier wave can be achieved by amplifying the carrier in a tuned amplifier whose collector load impedance is varied by the modulating signal. The principle of the method is illustrated by Fig. 3.20. When the modulating signal

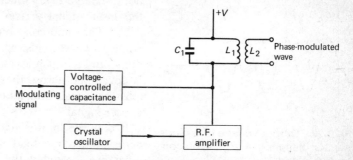

Fig. 3.20 Phase modulator

voltage is zero the collector tuned circuit C_1-L_1 is resonant at the crystal oscillator frequency. With the modulating signal applied, the total effective capacitance of the tuned circuit is varied and the circuit is de-tuned above and below resonance. This makes the phase of the output voltage alternately lag and lead the phase of the unmodulated output voltage, i.e. the output voltage is phase modulated.

Amplitude Demodulators

Demodulation or detection is the process of recovering the information carried by a modulated wave. The majority of d.s.b. amplitude-modulation sound broadcast receivers employ the diode detector, although increasingly the detection stage is included within an integrated circuit and then another kind of detector circuit is used. Many communication receivers operate with either s.s.b or i.s.b. signals and use some form of product detector.

The Diode Detector

Fig. 3.21 shows the circuit of a diode detector, consisting of a diode in series with a parallel resistor-capacitor network.

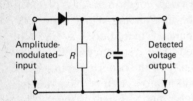

Fig. 3.21 Diode detector circuit

If an unmodulated carrier wave of constant amplitude is applied to the detector, the first positive half-cycle of the wave will cause the diode to conduct. The diode current will charge the capacitor to a voltage that is slightly less than the peak value of the input signal (slightly less because of a small voltage drop in the diode itself). At the end of this first half-cycle, the diode ceases to conduct and the capacitor starts to discharge through the load resistor R at a rate determined by the time constant, CR seconds, of the discharge circuit.

The time constant is chosen to ensure that the capacitor has not discharged very much before the next positive half-cycle of the input signal arrives to recharge the capacitor (see Fig. 3.22). The time constant for the charging of the capacitor is equal to Cr seconds, where r is the forward resistance of the diode and is much less than R. A nearly constant d.c. voltage is developed across the load resistor R; the fluctuations that exist are small and take place at the frequency of the input carrier signal.

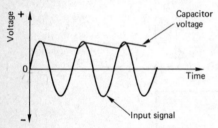

Fig. 3.22 Output voltage of a diode detector handling a signal of constant amplitude

If now the input signal is amplitude-modulated, the voltage across the diode load will vary in sympathy with the wave envelope, provided the time constant is small enough. The capacitor must be able to discharge rapidly enough for the voltage across it to follow those parts of the modulation cycle when the modulation envelope is decreasing in amplitude (see Fig. 3.23). The capacitor voltage falls until a positive half-cycle of the input signal makes the diode conduct and recharge the capacitor. When the modulation envelope is decreasing, one positive half-cycle is of lower peak value than the preceding positive half-cycle and the capacitor is recharged to a smaller

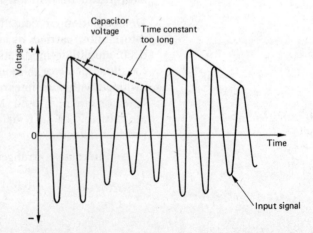

Fig. 3.23 Output voltage of a diode detector handling an amplitude-modulated signal

voltage. If the time constant of the discharge path is too long, relative to the periodic time of the modulating signal, the capacitor voltage will not be able to follow the troughs of the modulation envelope; that is, the decay curve passes right over the top of one or more input voltage peaks as shown by the dotted line in Fig. 3.23, and waveform distortion takes place.

The time constant must not be too short, however, or the voltage across the load resistor will not be as large as it could be, because insufficient charge will be stored between successive pulses of diode current. The time constant determines the rapidity with which the detected voltage can change, and must be long compared with the periodic time of the carrier wave and short compared with the periodic time of the modulating signal.

The voltage developed across the diode load resistor has three components: (a) a component at the wanted modulating signal frequency, (b) a d.c. component that is proportional to the peak value of the unmodulated wave (this component is not wanted for detection and must be prevented from reaching the following audio-frequency amplifier stage), and (c) components at the carrier frequency and harmonics of the carrier frequency that must also be prevented from reaching the audio-frequency amplifier. To eliminate the unwanted components the detector output is fed into a resistance-capacitance filter network before application to the audio-frequency amplifier.

Fig. 3.24 shows a possible filter circuit. Capacitor C_2 acts as a d.c. blocker to remove the d.c. component of the voltage appearing across load resistor R_1. Capacitor C_3 has a low reactance at the carrier frequency and its harmonics and, in conjunction with resistor R_2, filters out voltages at these frequencies. The voltage appearing across R_3 is therefore just the required modulating signal.

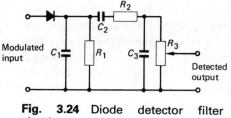

Fig. 3.24 Diode detector filter circuit

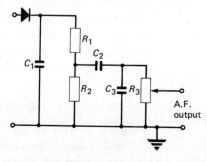

Fig. 3.25 Another diode detector filter circuit

An alternative arrangement is shown in Fig. 3.25. Capacitor C_2 is the d.c. blocker and C_3-R_3 is the r.f. filter; R_3 also functions as the volume control.

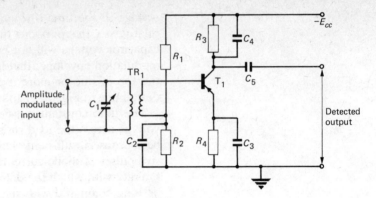

Fig. 3.26 The transistor detector

The Transistor Detector

The circuit of a common-emitter transistor detector is shown in Fig. 3.26. Rectification takes place in the emitter/base circuit and the rectified signal is amplified by the transistor in the usual way.

Components R_1, R_2, R_4, C_2 and C_3 provide bias and d.c. stabilization and R_3 is the collector load resistor. C_4 is a by-pass capacitor to prevent voltages at the carrier frequency appearing across R_3 and being fed, via C_5, to the output terminals of the circuit.

The input amplitude-modulated signal is fed into the base/emitter circuit via r.f. transformer TR_1, the primary winding of which is tuned to the carrier frequency. The base/emitter junction of transistor T_1 acts as a semiconductor diode and together with R_2 and C_2 forms a diode detector. The detected voltage appears across R_2 and varies the emitter/base bias voltage of the transistor. This variation causes the collector current to vary in accordance with the modulation envelope. A voltage at the modulating signal frequency appears across the collector load resistor R_3 and this is coupled to the load by capacitor C_5.

The disadvantage of the transistor detector is its limited dynamic range and for this reason it is not often used in broadcast receivers having discrete circuitry. The circuit is used in the detector section of some integrated circuits.

The Balanced Demodulator

Demodulation of an s.s.b or an i.s.b. signal can be achieved with any of the balanced modulators previously discussed. The carrier component which was suppressed at the transmitter must be re-inserted at the same frequency and the input or modulating signal is the transmitted sideband.

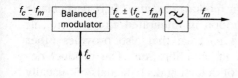

Fig. 3.27 Use of a balanced modulator as a demodulator

Suppose that the modulating signal is a sine wave of frequency f_m and that the lower sideband is transmitted. Then the input to the demodulator is at frequency $f_c - f_m$ and the demodulator output contains components at frequencies $f_c \pm (f_c - f_m)$. The lower sidefrequency is selected by a low-pass filter and is equal to $f_c - (f_c - f_m)$, or f_m, which is the required modulating signal (see Fig. 3.27).

The frequency of the re-inserted carrier must be very close to the frequency of the carrier which was suppressed at the transmitter. Otherwise the frequency components of the demodulated output waveform will bear the wrong relationships to one another. The maximum permissible frequency error depends upon the nature of the signal. It may be as large as ±15 Hz for speech transmissions but only 2 Hz for telegraphy/data signals. As in the case of the modulator it is essential for the carrier amplitude to be considerably larger than the signal amplitude to ensure that the diodes are switched by the carrier. When digital data is being received, it is necessary for the re-inserted carrier to also be phase-accurate.

Another method of demodulating an s.s.b. signal is known as the PRODUCT DETECTOR, two versions of which are shown in Fig. 3.28a and b. In both of these circuits the a.c. current conducted by the active device is proportional to the product of the s.s.b.s.c. and carrier voltages. The output current therefore contains components at a number of different frequencies amongst which is a component at the difference frequency. Assuming sinusoidal modulation this is equal to $f_c - (f_c - f_m)$ or f_m; thus the difference frequency is the required modulating signal. In both circuits, higher frequency unwanted components are filtered out by the shunt capacitor C_3.

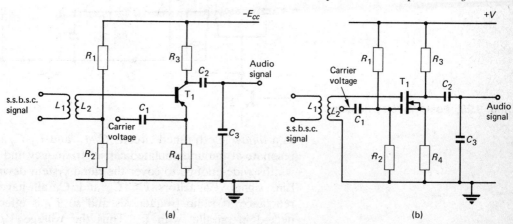

Fig. 3.28 Two product detectors

The product detector, also known as the heterodyne detector, is increasingly used in modern circuitry. It is particularly convenient for implementation in integrated circuit form, although normally within a package that also provides other circuit functions, such as i.f. amplification. The product demodulator is also capable of detecting d.s.b. signals but usually only the i.c. versions are used for this purpose.

Frequency Demodulators

The function of a frequency demodulator is to produce an output voltage whose magnitude is directly proportional to the frequency deviation of the input signal, and whose frequency is equal to the number of times per second the input signal frequency is varied about its mean value. Frequency demodulation can be achieved in several different ways. Most receivers using discrete circuitry employ either the ratio detector or the Foster–Seeley discriminator, but receivers incorporating integrated circuitry often use either the quadrature detector or a phase-locked loop.

The Foster–Seeley Discriminator

The circuit diagram of a Foster–Seeley discriminator is given in Fig. 3.29. The tuned circuit C_1-L_1 acts as the collector load for the final stage of the i.f. amplifier, which is generally operated

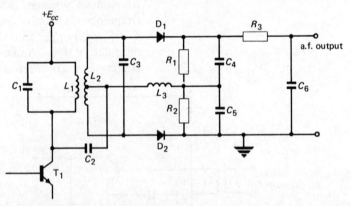

Fig. 3.29 Foster-Seeley detector

as a *limiter*. Both tuned circuits C_1-L_1 and L_2-C_3 are tuned to resonate at the unmodulated carrier frequency and have bandwidths wide enough to cover the rated system deviation of the f.m. signal. Capacitors C_2, C_4 and C_5 all have negligible reactance at radio frequencies and so L_1 is effectively connected in parallel with L_3. Thus the voltage V_1 developed across L_1 also appears across L_3.

Suppose the voltage appearing across L_1 is at the unmodulated carrier frequency. The current flowing in L_1 induces an

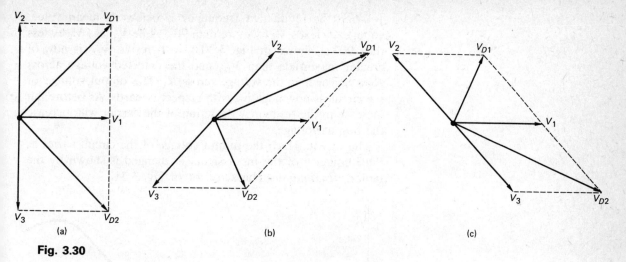

Fig. 3.30

e.m.f. into the secondary winding L_2 and this causes an in-phase current to flow in the series circuit L_2-C_3. A voltage is developed across C_3 which lags this current, and hence the induced voltage by 90°. Since inductor L_2 is accurately centre-tapped, one-half of this voltage appears across each half of L_2.

Let the voltage appearing across the upper half of the winding be labelled as V_2 with V_3 being the voltage across the lower half. The total voltages V_{D1} and V_{D2} applied across the diodes D_1 and D_2 are, respectively, the phasor sums of the voltages V_1 and V_2, and V_1 and V_3. The phase relationships are such that V_2 leads V_1 by 90° and V_3 lags V_1 by 90° as shown by the phasor diagram of Fig. 3.30a. Since $V_{D1} = V_{D2}$, equal amplitude detected voltages appear across the diode load resistors R_1 and R_2. Because of the diode connections, these two voltages act in opposite directions and cancel out, so that the voltage appearing across the output terminals of the circuit is zero.

When the frequency of the signal voltage developed across L_1 is above the unmodulated carrier frequency, the voltage across C_3 will lead the e.m.f. induced into L_2 by some angle greater than 90°. This results in V_2 leading V_1 by an angle less than 90° and V_3 lagging V_1 by more than 90° (Fig. 3.30b). Now the voltage V_{D1} applied across diode D_1 is larger than the voltage V_{D2} applied to D_2 and so the voltage developed across load resistor R_1 is greater than the voltage across R_2. A positive voltage, equal to the difference between the two load voltages, is produced at the output terminals. If the frequency deviation of the carrier is increased, the larger will become the difference between the magnitudes of the diode voltages V_{D1} and V_{D2}, and the output voltage will increase in the positive direction.

When the modulated frequency is below its mean value, voltage V_2 leads V_1 by more than 90°, while V_3 lags V_1 by less than 90° as shown in Fig. 3.30c. As a result, V_{D2} is now of greater magnitude than V_{D1} and the detected voltage across R_2 is bigger than the voltage across R_1. The output voltage of the circuit is now negative with respect to earth. As before, an increase in the frequency deviation of the carrier will increase the output voltage.

The way in which the output voltage of the circuit varies as the frequency of the input signal is changed is shown by the typical *discriminator* characteristic of Fig. 3.31.

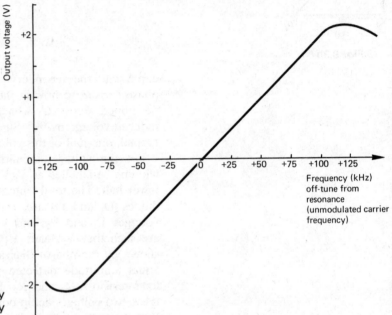

Fig. 3.31 Output voltage/frequency characteristic of a Foster-Seeley detector

Operation of the detector should be restricted to the linear part of the characteristic. The *turn-over* points are produced by the limited bandwidth of the tuned circuits C_1-L_1 and C_3-L_2, reducing the voltages applied to the diodes.

The output voltage of the circuit will also vary if the amplitude of the input signal should vary. This is, of course, an undesirable effect and to prevent it happening the detector should be preceeded by one or more stages of amplitude limiting. De-emphasis of the output signal is provided by R_3 and C_6.

The Ratio Detector

A commonly used f.m. detector, particularly for broadcast receivers, is the ratio detector, one form of which is given by Fig. 3.32. The main advantage of this circuit over the Foster-Seeley detector is that it incorporates its own amplitude limiting action and often a separate limiter is not needed.

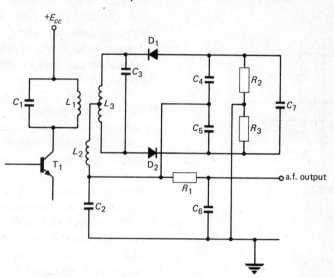

Fig. 3.32 The ratio detector

Inductor L_1 is inductively coupled to both L_2 and L_3 but L_2 and L_3 are not coupled together. The tuned circuits C_1-L_1 and C_3-L_3 are each tuned to the unmodulated carrier frequency. When a voltage at this frequency appears across L_1, voltages are induced into both L_2 and L_3. Capacitors C_2, C_4 and C_5 have negligible reactance at radio frequencies and so the voltage applied across diode D_1 is the phasor sum of the voltages across L_2 and the upper half of L_3. Similarly the voltage applied to D_2 is the phasor sum of the voltages across L_2 and the lower half of L_3. If the voltage across L_2 is labelled as V_1 and the other two voltages are labelled V_2 and V_3 respectively, the phasor diagram given in Fig. 3.30 will represent the voltages. The resultant voltages V_{D1} and V_{D2} applied to the diodes will vary with frequency to produce voltages across the load capacitors C_4 and C_5. The voltage across the *d.c. load capacitor* C_7 is the sum of the voltages across C_4 and C_5 and, since $R_2 = R_3$, one-half of this voltage appears across each resistor. The time constant $C_7(R_2 + R_3)$ is sufficiently long to ensure that the voltage across C_7 remains more or less constant at very nearly the peak voltage appearing across C_3.

The *audio load capacitor* C_2 is connected between the junctions of C_4/C_5 and R_2/R_3, and this part of the circuit is re-drawn in Fig. 3.33. The voltages across C_5 and R_3 have the

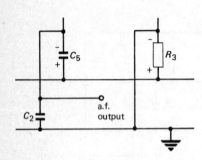

Fig. 3.33

polarities shown and, when the input signal is at the unmodulated carrier frequency, are of equal magnitude. The two voltages act in opposite directions, with the result that no current flows, and the voltage across the audio load capacitor C_2 is zero. When the input signal frequency increases, the voltage applied to diode D_1 increases and the voltage across D_2 falls. As a result the voltage across C_4 increases while the voltage across C_5 falls; but, since the *sum* of these voltages remains constant, the voltage across R_3 does not change. A current now flows in the circuit of Fig. 3.33 and a positive voltage, equal to the difference between V_{C5} and V_{R3}, appears across C_2. If the frequency deviation is increased, the voltage across C_5 will fall still further and the voltage across C_2 will increase. Conversely, if the input frequency is reduced, the voltage across C_5 will become bigger than the voltage across R_3, and a current will flow in the opposite direction to before to produce a negative output voltage. When the frequency of the input signal voltage is modulated, the modulating signal voltage will appear across C_2. Components R_1 and C_6 provide de-emphasis of the output voltage.

The output-voltage/input-frequency characteristic of a ratio detector has the same shape as the Foster-Seeley curve shown in Fig. 3.31. However, the output voltage available for a given frequency deviation is only one-half that provided by the Foster–Seeley circuit, and the linearity of the characteristic is not as good. The advantage of the ratio detector has been previously mentioned; it provides some degree of self-limiting in the following manner. If the amplitude of the input signal is steady, the voltage across the d.c. load capacitor C_7 is constant because of the long time constant $C_7 (R_2 + R_3)$. An increase in the input signal voltage will cause both diodes to conduct extra current, and this results in an increased volts drop across the tuned circuit, tending to keep the voltage applied to the diodes more or less constant. A similar action takes place if the input signal voltage should fall; the diodes pass a smaller current and the volts drop across the tuned circuit is reduced, allowing the diode voltage to rise. The variable voltage drop across the tuned circuit in conjunction with the long time constant of the diode load ensure that the output voltage responds very little, if at all, to any changes in input signal amplitude.

The ratio detector shown in Fig 3.32 is *balanced* since the d.c. voltage appearing across the d.c. load capacitor C_7 is balanced with respect to earth potential. Sometimes it is convenient to have an unbalanced voltage and then an unbalanced circuit is used. Fig. 3.34 shows the circuit of one version of the unbalanced ratio detector; its operation is left as an exercise for the reader (see Exercise 3.14).

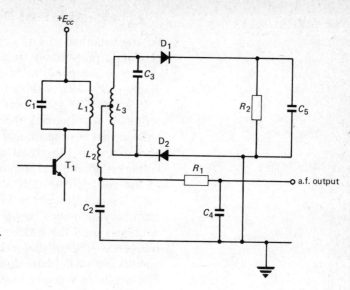

Fig. 3.34 Unbalanced ratio detector

The Quadrature Detector

The action of a quadrature detector depends upon two voltages, which are both derived from the f.m. signal to be detected and which are 90° out of phase with one another at the unmodulated carrier frequency. Essentially, the detector compares the relative phases of the signal on either side of a single tuned circuit. Quadrature detection is rarely found in discrete form but is increasingly used in modern equipment since it is convenient for fabrication within an integrated circuit. Generally the i.c. includes an i.f. amplifier, a limiter, the quadrature detector, and, sometimes, an audio-frequency pre-amplifier. Two rectangular voltages, both derived from the f.m. signal to be demodulated, are applied to a transistor circuit which produces an output voltage only when *both* input voltages are present. At the unmodulated carrier frequency, the two input voltages have a time difference of a quarter of a period between them, corresponding to the 90° (quadrature) phase shift (Fig. 3.35a). When the frequency of the f.m. signal is above its umodulated value, voltage waveform *b* is advanced in time (Fig. 3.35b) and the pulses of output voltage occupy longer intervals of time. Conversely when the signal frequency is decreased, the pulses of output voltage are narrower (Fig. 3.35c). The *mean* value of the output voltage is thus proportional to the frequency deviation of the input signal voltage. If the input signal is frequency modulated, the mean value of the output voltage will vary with the waveform of the modulating signal.

The quadrature detector is often used in i.c. form since all the necessary components except one can be fabricated within the chip. The exception is the *quadrature coil* which is required to produce the necessary time difference between voltages A and B.

The Phase-locked Loop Detector

Another method of f.m. detection which has only become an economic proposition since the advent of linear integrated circuits is the phase-locked loop (p.l.l.), the block schematic diagram of which is shown in Fig. 3.36.

If a signal at a constant frequency is applied to the input terminals of the circuit, the phase detector produces an output voltage that is proportional to the instantaneous *phase* difference between the signal and oscillator voltages. The *error* voltage is filtered and amplified before it is applied to the input of the voltage-controlled oscillator. The error voltage varies the oscillator frequency in the direction which reduces the frequency difference between signal and oscillator. This action continues until the oscillator frequency is equal to the signal frequency. The oscillator is then said to be *locked*; in this condition a small *phase* difference will exist between the signal and oscillator voltages in order to generate the error voltage needed to maintain the lock.

If the input signal frequency should change, the error voltage will change also, with the appropriate polarity, and force the oscillator frequency to follow. When the input signal is frequency modulated, the error voltage will vary in the same way as the required modulating signal and so the circuit acts as an f.m. demodulator. The circuit can be made using discrete components or using integrated operational amplifiers but most convenient is the integrated p.l.l. circuit.

Fig. 3.37 shows the basic circuit of an integrated p.l.l. f.m. demodulator. C_1 and C_2 couple the previous stage in the radio receiver to the detector and C_3 is the tuning capacitor of the voltage-controlled oscillator. C_5 is a part of the low-pass filter between the phase detector and the amplifier, and lastly C_4 and R_1 are the de-emphasis components.

The p.l.l. detector offers a number of advantages over its competitors, namely: (*a*) the detector is tuned to the unmodulated carrier frequency by a single external capacitor; (*b*) the upper-frequency limit is high; (*c*) it introduces little noise or distortion; and (*d*) it does not require an inductance. The circuitry is complicated and not competitive economically if discrete components are used. The integrated circuit versions also used to be expensive but their present cost is such that they are employed in some communication receivers.

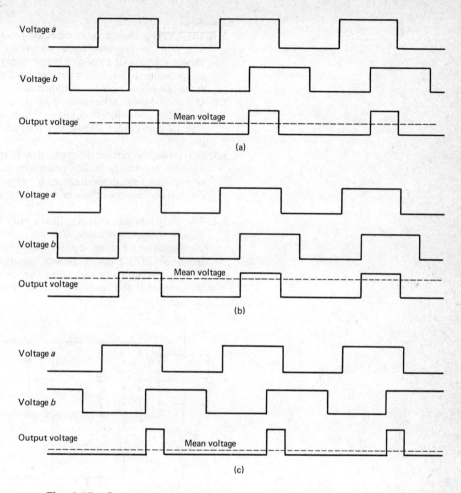

Fig. 3.35 Operation of a quadrature detector (Note: mean voltage not drawn to scale)

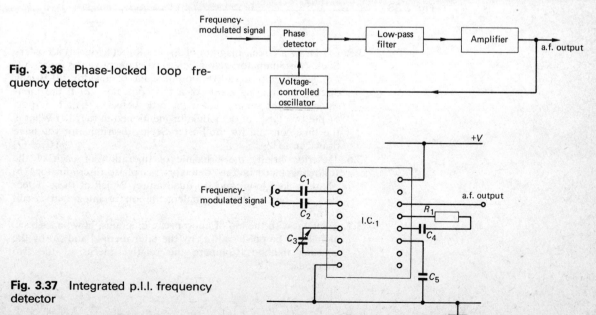

Fig. 3.36 Phase-locked loop frequency detector

Fig. 3.37 Integrated p.l.l. frequency detector

Exercises

3.1. (*a*) Explain how a ratio detector produces amplitude limiting. (*b*) With severe amplitude variations of the input signal the limiting action of a ratio detector may produce distortion of the output signal. Explain what causes this type of distortion and how the effect may be minimized. (*C & G*)

3.2. Draw a block schematic diagram of a phase-locked loop frequency-modulation detector. Explain the principle of operation of the circuit. What are the advantages of this type of detector?

3.3. (*a*) Draw the circuit diagram of a Foster–Seeley discriminator with an amplitude limiter preceding it. (*b*) Describe its operation as an f.m. demodulator. (*c*) Which parts of your circuit control the bandwidth over which the discriminator operates? (*C & G*)

3.4. Fig. 3.38 shows a discriminator. (*a*) What type is it? Draw phasor diagrams relating V_1 with each of V_2, V_3 and V_4 when the signal is (i) at the carrier frequency, (ii) above the carrier frequency. (*c*) Draw a typical input/output characteristic for such a discriminator, labelling the axes. (*d*) What happens to the output if the incoming carrier drifts from its nominal frequency? (*C & G*)

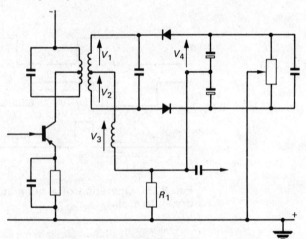

Fig. 3.38

3.5. (*a*) Draw a circuit diagram of (i) a ratio detector, (ii) a Foster–Seeley discriminator, giving typical component values for a v.h.f. f.m. system with a 15 kHz output signal bandwidth. (*b*) Briefly discuss the limiting action of a ratio detector. (*c*) If the input signals are the same, what is the ratio between the a.f. outputs of the two types of demodulator mentioned in (*a*)? (*d*) What is the time constant for the Foster–Seeley discriminator you have drawn in (*a*)? (*C & G*)

3.6. Describe briefly the principle of operation of each of the following kinds of f.m. detector: (*a*) phase discriminator, (*b*) phase-locked loop, and (*c*) quadrature. Which of these detectors are best suited to implementation in integrated circuit form?

3.7. Explain, with the aid of appropriate diagrams, how an s.s.b.s.c. signal can be produced (*a*) by the filter method and (*b*) by the phasing method. Compare the relative merits of the two methods.

3.8. Draw the circuit diagram of a diode detector suitable for the demodulation of d.s.b. amplitude-modulation signals. Explain the operation of the circuit and give typical component values.

3.9. (*a*) What is a varactor diode? (*b*) Draw the circuit diagram of a varactor diode modulator and explain its operation. (*c*) At what stage in an f.m. transmitter is pre-emphasis applied? (*d*) Sketch a typical pre-emphasis characteristic, annotating the axes.

(*C & G*)

3.10. Describe with circuit diagrams the function and operation in frequency modulation equipment of (*a*) a limiter, (*b*) a reactance modulator. (*C & G*)

3.11. (*a*) Explain how the input-voltage/output-current characteristic of a semiconductor diode is used to demodulate an amplitude-modulated wave. (*b*) An amplitude-modulated signal is applied to the input of the diode detector shown in Fig. 3.24. By reference to current or voltage waveforms describe the function of each component in reproducing the modulating signal.

(*C & G*)

3.12. Fig. 3.39 shows the pin connections of an integrated circuit double-balanced modulator. The pin functions are as listed:

1 + signal in   8 + power supply voltage
2 − signal in   10 + output
3 + carrier in   12 − output
4 decouple   13 carrier leak adjust
7 + carrier in   14 earth.

(*a*) Draw a suitable modulator circuit using this i.c. (*b*) List the advantages of i.c.s over the use of discrete components.

3.13. Explain, with the aid of block diagrams, how a frequency modulated wave can be produced using (*a*) a direct method and (*b*) an indirect method. Why is the indirect method employed?

3.14. Explain, with the aid of phasor diagrams, the operation of the ratio detector circuit given in Fig. 3.34.

Short Exercises

3.15. Draw the circuit diagram of a balanced modulator that employs two junction field-effect transistors as the non-linear elements.

3.16. Draw the block diagram of the phase-shifting method of producing an s.s.b.s.c. signal in which the upper sideband is transmitted.

3.17. Should the bandwidth of the tuned circuits of a Foster–Seeley discriminator be greater or less than the rated system deviation of the signal to be detected? Give a reason for your answer.

3.18. Draw the circuit diagram of a double-balanced demodulator using an integrated circuit.

3.19. (*a*) What is the function of the modulator stage in a radio transmitter? (*b*) What is the function of the detector stage in a radio receiver?

3.20. What are the requirements for the demodulation of an s.s.b.s.c. signal in terms of (*a*) the magnitude of the re-inserted carrier, (*b*) the frequency of the re-inserted carrier, and (*c*) the phase of the re-inserted carrier?

3.21. Draw the circuit diagram of a reactance frequency modulator using a field-effect transistor. State which kind of f.e.t. you have drawn.

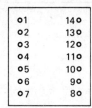

Fig. 3.39

3.22. List the component frequencies which appear across the load resistor of a diode detector. Explain, with the aid of a diagram, how the unwanted components are removed before the detected signal is applied to the audio amplifier.

3.23. A 2 kHz sinusoidal signal is applied to a balanced modulator along with a 80 kHz carrier wave and the lower sidefrequency is selected by a suitable filter. At the receiver the s.s.b.s.c. signal is applied to a balanced demodulator together with a carrier at 80.045 kHz. Determine the frequency of the demodulated signal.

3.24. Fig. 3.40 shows the circuit of an integrated product detector. Suggest the function of each component shown.

3.25. List the relative advantages and disadvantages of
(*a*) a Foster–Seeley discriminator, a ratio detector, and a quadrature detector.
(*b*) a diode detector, a transistor detector, and a product detector.

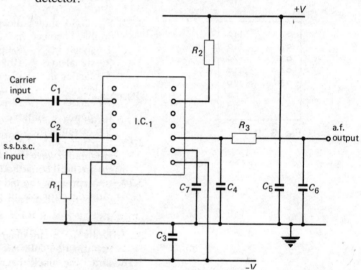

Fig. 3.40

4 Transmission Lines

Introduction

The basic purpose of a transmission line is to transmit electrical energy from one point to another. The length of a line may be several kilometres in the case of a line communication system, some tens or hundreds of metres in the case of a *feeder* used to connect a radio transmitter or receiver to its aerial, or perhaps only a fraction of a metre when the line is used as an integral part of a u.h.f. equipment.

Essentially a transmission line consists of a pair of conductors separated from one another by a dielectric. The two main types of line used in radio communication systems are the two-wire or twin line and the coaxial pair. Both types of cable are available with air as the dielectric, or with some insulating material such as polythene. Each conductor of a pair has both series resistance and inductance, and shunt capacitance and leakance exists between the conductors. The magnitudes of the four *primary coefficients* depend upon the physical dimensions of the conductors and the nature of the dielectric used. The values of the resistance and the leakance also depend upon the frequency of the signal propagating along the line.

Matched Transmission Lines

The behaviour of a transmission line when a signal is applied across its input terminals is determined by its *secondary coefficients*. The four secondary coefficients of a line are its characteristic impedance, its attenuation and phase-change coefficients, and its phase velocity of propagation.

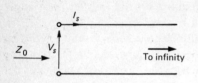

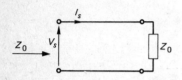

Fig. 4.1 Characteristic impedance of a line

Characteristic Impedance

The characteristic impedance Z_0 of a transmission line is the input impedance V_S/I_S of the line when either the line is of infinite length or it is terminated in its characteristic impedance. The two definitions, illustrated by Fig. 4.1, are essentially the same as demonstrated in TSII.

The characteristic impedance of a radio-frequency line is determined by the inductance and capacitance of the line in accordance with equation (4.1):

$$Z_0 = \sqrt{(L/C)} \text{ ohms} \tag{4.1}$$

The characteristic impedance of a particular r.f. line is of constant magnitude, since neither the inductance L nor the capacitance C varies with frequency, and is purely resistive. This means that at all points along the line the current and voltage are in phase with one another.

In turn, the inductance and capacitance of a line depend upon its dimensions, such as conductor diameter and spacing and so therefore does the characteristic impedance.

For a two-wire line

$$Z_0 = \frac{276}{\sqrt{\varepsilon_r}} \log_{10} \frac{D}{r} \text{ ohms} \tag{4.2}$$

where D is the spacing between the centres of the two conductors, r is the radius of each conductor, and ε_r is the relative permittivity of the continuous dielectric.

For a coaxial line

$$Z_0 = \frac{138}{\sqrt{\varepsilon_r}} \log_{10} \frac{R}{r} \text{ ohms} \tag{4.3}$$

where R is the inner radius of the outer conductor and r is the radius of the inner conductor.

Because of constructional difficulties, such as the need to avoid very small conductor spacings, not all characteristic impedance values are available. For two-wire lines, Z_0 is normally bigger than $100 \, \Omega$ and for coaxial lines Z_0 is usually 30–$100 \, \Omega$.

Attenuation Coefficient

As a current or voltage wave is propagated along a line, its amplitude is progressively reduced or *attenuated*, because of losses in the line. These losses are of three types: firstly there are conductor losses caused by I^2R power dissipation in the series resistance of the line; secondly there are dielectric losses; and thirdly, radiation losses. Radiation losses will occur in a balanced twin line at frequencies high enough for the conductor separation to be an appreciable fraction of the signal wavelength. Radiation loss does not occur in a coaxial pair in which the outer is made of solid copper, but some losses will take place when a braided outer is used.

The radiation losses are difficult to determine but should be small if the correct type of cable is used. Neglecting the radiation losses, it is found that, if the current or voltage at the sending-end of the line is I_s or V_s, then the current or voltage one metre along the line is

$$I_1 = I_s e^{-\alpha} \quad \text{or} \quad V_1 = V_s e^{-\alpha}$$

where e is the base of natural logarithms (2.7183) and α is the ATTENUATION COEFFICIENT of the line in nepers per metre, where 1 neper = 8.686 dB. In the next metre length of line, the attenuation is the same and so the current I_2, two metres along the line, is

$$I_2 = I_1 e^{-\alpha} = I_s e^{-\alpha} e^{-\alpha} = I_s e^{-2\alpha}$$

Similarly $V_2 = V_s e^{-2\alpha}$

If the line is l metres in length, the current and voltage received at the outputs terminals of the line are given, respectively, by

$$I_r = I_s e^{-\alpha l} \tag{4.4}$$

$$V_r = V_s e^{-\alpha l} \tag{4.5}$$

Equations (4.4) and (4.5) show that both the current and voltage waves decay exponentially as they propagate along the length of the line.

At radio frequencies the attenuation coefficient of a line is given by equation (4.6)

$$\alpha = R/2Z_0 + GZ_0/2 \text{ nepers/metre} \tag{4.6}$$

where R is the series resistance per metre loop and G is the leakance per metre.

The attenuation coefficient is not a constant quantity but increases with increase in frequency. The two contributory parts of the attenuation coefficient vary with frequency in different ways; the conductor losses are proportional to the square root of frequency but the dielectric losses are directly proportional to frequency. Normally, the conductor losses are several times larger than the dielectric losses and often, particularly with coaxial lines, the dielectric loss is small enough to be neglected. Then the line attenuation is proportional to the square root of frequency, while the signal wavelength is inversely proportional to frequency. This means that the attenuation per wavelength *decreases* with increase in frequency. Often, particularly at the v.h.f. and higher bands, the loss of a line is small enough to be neglected; then the line is usually described as being *low-loss* or *loss-free*.

Phase Change Coefficient

As a current or voltage wave travels along a line, it experiences a progressive phase lag relative to its phase at the sending-end of the line. The PHASE-CHANGE COEFFICIENT β of a line is the number of radians or degrees phase lag per metre. If, for example, $\beta = 2°$ per metre, then a 10 metre length would introduce a phase shift of 20°. At radio frequencies the phase change coefficient is given by equation (4.7):

$$\beta = \omega\sqrt{(LC)} \text{ radians/metre} \tag{4.7}$$

Clearly the phase change coefficient is directly proportional to frequency. At the higher frequencies, lines are often quoted in terms of their *electrical length*; this is the product βl expressed in wavelengths.

EXAMPLE 4.1

A transmission line has a phase change coefficient β of 30° per metre at a particular frequency. If the physical length of the line is 1.5 m calculate its electrical length.

Solution

$$\beta l = 30° \times 1.5 = 45°$$

In one wavelength a phase shift of 360° takes place. Therefore,

$$\text{Electrical length} = \frac{45}{360}\lambda = \frac{\lambda}{8} \quad (Ans.)$$

Phase Velocity of Propagation

The phase velocity of propagation V_p of a transmission line is the velocity with which a sinusoidal current or voltage wave is propagated along a line. Any sinusoidal wave travels with a velocity of one wavelength per cycle and since there are f cycles per second this corresponds to a velocity of λf metres per second. Therefore,

$$V_p = \lambda f \text{ metres/second} \tag{4.8}$$

where λ is the wavelength and f is the frequency of the sinusoidal wave.

In a distance of one wavelength a phase lag of 2π radians occurs and so

$$\beta = 2\pi/\lambda \text{ radians}$$

Therefore,

$$\lambda = 2\pi/\beta$$

and

$$V_p = \frac{2\pi}{\beta} \cdot f = \omega/\beta \qquad (4.9)$$

At radio-frequencies

$$V_p = \omega/\omega\sqrt{(LC)} = 1/\sqrt{(LC)} \text{ metres/second} \qquad (4.10)$$

and has the same value at all frequencies. This means that all the component frequencies of a complex signal will propagate along a line at the same velocity and arrive at the end of the line together. Thus, the signal envelope will not suffer group delay/frequency distortion [TSII].

The phase velocity of propagation on a line is always somewhat smaller than the velocity of light $(c = 3 \times 10^8 \text{ m/s})$. Usually, the velocity is somewhere between $0.6\,c$ and $0.9\,c$.

Propagation on a Matched Line

The input impedance of a matched line, that is one that is terminated in Z_0, is equal to Z_0. Suppose a generator of e.m.f. E_s and impedance Z_s is connected across the input terminals of the line. The voltage appearing across the terminals is then

$$V_s = E_s Z_0/(Z_s + Z_0)$$

and the input current I_s is equal to

$$E_s/(Z_s + Z_0)$$

The input current and voltage propagate along the line and are attenuated and phase shifted as they travel. At any point along the line the ratio of the voltage and the current at that point is equal to the characteristic impedance Z_0. At the receiving end of the line, *all* the power carried by the waves is dissipated in the load impedance.

Mismatched Transmission Lines

Very often a transmission line is operated with a terminating impedance that is not equal to the characteristic impedance of the line. Sometimes this may be intentional but most often it is simply because the correct terminating impedance is not possible, or is not available. In a radio system the terminating impedance is usually an aerial. This will have an impedance which depends upon the type of aerial and it is not always convenient to use a line having the same value of characteristic impedance.

When the impedance terminating a line is not equal to the characteristic impedance, the line is said to be *incorrectly terminated* or *mismatched*. Since the line is not matched, the

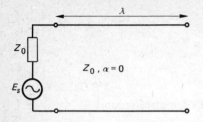

Fig. 4.2 Open-circuited loss-free line

load is not able to absorb all the power incident upon it and so some of the power is *reflected* back towards the sending end of the line.

Consider Fig. 4.2 which shows a loss-free line whose output terminals are open circuited. The line has an electrical length of one wavelength and its input terminals are connected to a source of e.m.f. E_s volts and impedance Z_0 ohms.

When the generator is first connected to the line, the input impedance of the line is equal to its characteristic impedance Z_0. An *incident* current of $E_s/2Z_0$ then flows into the line and an *incident* voltage of $E_s/2$ appears across the input terminals. These are, of course, the same values of sending-end current and voltage that flow into a correctly-terminated line. The incident current and voltage waves propagate along the line, being phase-shifted as they travel. Since the electrical length of the line is one wavelength, the overall phase shift experienced is 360°.

Since the output terminals of the line are open-circuited, no current can flow between them. This means that *all* of the incident current must be *reflected* at the open-circuit. The total current at the open-circuit is the phasor sum of the incident and reflected currents, and since this must be zero the current must be reflected with 180° phase shift. The incident voltage is also totally reflected at the open circuit but with zero phase shift. The total voltage across the open-circuited terminals is twice the voltage that would exist if the line were correctly terminated. The reflected current and voltage waves propagate along the line towards its sending-end, being phase-shifted as they go. When the reflected waves reach the sending end, they are completely absorbed by the impedance of the matched source.

At any point along the line, the total current and voltage is the phasor sum of the incident and reflected currents and voltages. Consider Fig. 4.3. At the open-circuit the phasors representing the incident and reflected currents are of equal length (since *all* the incident current is reflected) and point in opposite directions. The current flowing in the open circuit is the sum of these two phasors and is therefore zero as expected. At a distance of $\lambda/8$ from the open circuit, the incident current phasor is 45° leading, and the reflected current phasor is 45° lagging on the open-circuit phasors. The lengths of the two phasors are equal since the line loss is zero but they are 90° out of phase with one another. The total current at this point is $\sqrt{2}$ times the incident current. Moving a further $\lambda/8$ along the line, the incident and reflected current phasors have rotated, in opposite directions, through another 45° and are now in phase with one another. The total current $\lambda/4$ from the open circuit is equal to twice the incident current. A further

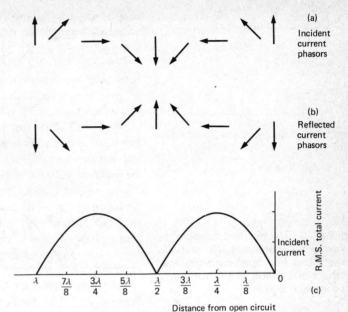

(a)
Incident
current
phasors

(b)
Reflected
current
phasors

Fig. 4.3 (a) The incident current, (b) the reflected current at λ/8 intervals along a loss-free open-circuited line, (c) the r.m.s. value of the total current at each point

Incident current

R.M.S. total current

λ $\dfrac{7\lambda}{8}$ $\dfrac{3\lambda}{4}$ $\dfrac{5\lambda}{8}$ $\dfrac{\lambda}{2}$ $\dfrac{3\lambda}{8}$ $\dfrac{\lambda}{4}$ $\dfrac{\lambda}{8}$ 0

(c)

Distance from open circuit

λ/8 along the line finds the two phasors once again at right angles to one another so that the total line current is again √2 incident current. At a point λ/2 from the end of the line, the incident and reflected current phasors are in antiphase with one another and the total line current is zero. Over the next half-wavelength of line the phasors continue to rotate in opposite directions, by 45° in each λ/8 distance, and the total line current is again determined by their phasor sum.

It is usual to consider the r.m.s. values of the total line current and then its phase need not be considered. The way in which the r.m.s. line current varies with distance from the open-circuit is shown by Fig. 4.3c. The points at which maxima (*antinodes*) and minima (*nodes*) of current occur are always the same and do not vary with time. Because of this the waveform of Fig. 4.3c is said to be a STANDING WAVE.

If, now, the voltages existing on the line of Fig. 4.2 are considered, the phasors shown in Fig. 4.4 are obtained. At the open-circuited output terminals, the incident and reflected voltage phasors are in phase and the total voltage is twice the incident voltage. Moving from the open circuit towards the sending end of the line, the phasors rotate through an angle of 45° in each λ/8 length of line; the incident voltage phasors rotate in the anti-clockwise direction and the reflected voltage phasors rotate clockwise. The total voltage at any point along the line is the phasor sum of the incident and reflected voltages and its r.m.s. value varies in the manner shown in Fig. 4.4c.

Two things should be noted from Figs. 4.3c and Fig. 4.4c. Firstly, the *voltage standing-wave* pattern is displaced by λ/4

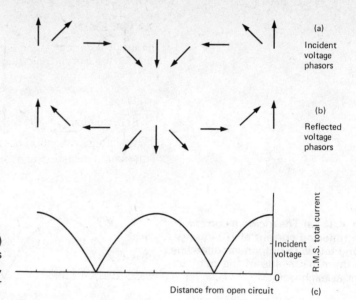

(a)
Incident
voltage
phasors

(b)
Reflected
voltage
phasors

Fig. 4.4 (a) The incident voltage, (b) the reflected voltage at λ/8 intervals along a loss-free open-circuited line, (c) the r.m.s. value of the total voltage at each point

from the *current standing-wave* pattern, i.e. a current antinode occurs at the same point as a voltage node and vice versa. Secondly, the current and voltage values at the open circuit are repeated at λ/2 intervals along the length of the line; this remains true for any longer length of loss-free line.

When the output terminals of a loss-free line are short circuited, the conditions at the termination are reversed. There can be no voltage across the output terminals but the current flowing is twice the current that would flow in a matched load. This means that at the short circuit the incident current is totally reflected with zero phase shift and the incident voltage is totally reflected with 180° phase shift. Thus, Fig. 4.3c shows how the r.m.s. voltage on a short-circuited line varies with distance from the load, and Fig. 4.4c shows how the r.m.s. current varies.

Clearly, neither an open-circuited nor a short-circuited line can be used for the transmission of information from one point to another. Either condition might arise because of a fault but might be intended, particularly the short-circuit, when the line is to be used to simulate an electrical component.

Open and short-circuit terminations are the two extreme cases of a mismatched line and in most cases the mismatched load will have an impedance somewhere in between. The fraction of the incident current or voltage that is reflected by the load is determined by the REFLECTION COEFFICIENT of the load. The *voltage reflection coefficient* ρ_v is the ratio reflected-voltage/incident-voltage, and the *current reflection coefficient* ρ_i is the ratio reflected-current/incident-current, at the load. Always $\rho_i = -\rho_v$.

EXAMPLE 4.2

The incident current at the output terminals of a mismatched line is 5 mA. If the current reflection coefficient of the load is 0.5, what is the reflected current?

Solution

Reflected current $= 0.5 \times 5 = 2.5$ mA   (*Ans.*)

The voltage reflection coefficient is determined by the values of the characteristic impedance of the line and the load impedance. Thus,

$$\rho_v = \frac{Z_L - Z_0}{Z_L + Z_0} \qquad (4.11)$$

The magnitude of the voltage reflection coefficient lies in the range ± 1, the limits corresponding to the cases of open and short-circuited loads.

EXAMPLE 4.3

Calculate the voltage reflection coefficient of a line of 50 Ω characteristic impedance terminated by an impedance of (i) 50 Ω, (ii) 30 Ω, (iii) 100 Ω

Solution
(i) From equation (4.11)

$$\rho_v = \frac{50 - 50}{50 + 50} = 0 \qquad (Ans.)$$

This answer is to be expected since, if the line is matched, $Z_0 = Z_L$ and there is no reflection.

(ii) $\rho_v = \dfrac{30 - 50}{30 + 50} = -0.25 = 0.25\underline{/180°}$ (*Ans.*)

(iii) $\rho_v = \dfrac{100 - 50}{100 + 50} = 0.33 = 0.33\underline{/0°}$ (*Ans.*)

When the reflection coefficient is less than unity, the reflected current and voltage at any point along a loss-free line will be smaller than the incident values. Then the maximum current or voltage on the line will be less then twice the incident current and voltage, and the minimum current or voltage will not be zero. Suppose, for example, that $\rho_v = 0.5$ $\underline{/0°}$. Then the maximum line voltage will be 1.5 times the incident voltage and will occur at the load and at multiples of $\lambda/2$ from the load. The minimum line voltage will be 0.5 times the incident voltage and will occur $\lambda/4$ from the load and then at multiples of $\lambda/2$ from that point. Similarly, the maximum line current will be 1.5 times and the minimum line current will be 0.5 times the incident current. Maxima of voltage will occur at the same points as minima of current and vice versa. Fig. 4.5 shows the standing waves of current and voltage on a loss-free line with a load voltage reflection coefficient of 0.5 $\underline{/0°}$.

Fig. 4.5 Standing wave pattern on a mismatched line with $\rho_v = 0.5 \,\underline{/0^\circ}$

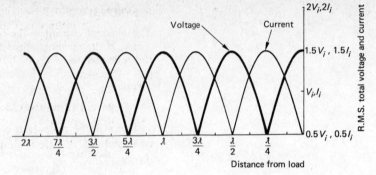

Distance from load

Standing Wave Ratio

An important parameter of any mismatched low-loss transmission line is its VOLTAGE STANDING-WAVE RATIO or v.s.w.r. The v.s.w.r. is the ratio of the maximum voltage on the line to the minimum voltage, i.e.

$$S = V_{max}/V_{min} \tag{4.12}$$

The maximum voltage on a mismatched line occurs at those points where the incident and the reflected voltages are in phase with one another. Also, the minimum line voltage exists at those points along the line at which the incident and reflected voltages are in antiphase. Therefore

$$S = \frac{V_i + |\rho_v| V_i}{V_i - |\rho_v| V_i} = \frac{1 + |\rho_v|}{1 - |\rho_v|} \tag{4.13}$$

where V_i is the incident voltage.

EXAMPLE 4.4

A low-loss line whose characteristic impedance is 70 Ω is terminated by an aerial of 75 Ω impedance. Determine the v.s.w.r. on the line.

Solution
From equation (4.11)

$$\rho_v = \frac{75 - 70}{75 + 70} = 0.035 \,\underline{/0^\circ}$$

Therefore, from equation (4.12)

$$S = \frac{1 + 0.035}{1 - 0.035} = 1.07 \quad (Ans.)$$

The presence of a standing wave on an aerial feeder is undesirable for several reasons and very often measures, beyond the scope of this book, are taken to approach the matched condition and to minimize reflections. The reasons why standing waves on a feeder should be avoided if at all possible are as follows:

(*a*) Maximum power is transferred from a transmission line to its load when the load impedance is equal to the characteristic impedance. When a load mismatch exists, some of the incident power is reflected at the load and the transfer efficiency is reduced.

(*b*) The power reflected by a mismatched load will propagate, in the form of current and voltage waves, along the line towards its sending end. The waves will be attenuated as they travel and so the total line loss is increased.

(*c*) At a voltage maximum the line voltage may be anything up to twice as great as the incident voltage. For low-power feeders, such as those used in conjunction with radio receivers, the increased voltage will not matter. For a feeder connecting a high power radio transmitter to an aerial, however, the situation is quite different. Care must be taken to ensure that the maximum line voltage will not approach the breakdown voltage of the line's insulation. This means that for any given value of v.s.w.r. there is a corresponding peak value for the incident voltage and hence for the maximum power that the feeder is able to transmit. A high v.s.w.r. on a feeder can result in dangerously high voltages appearing at the voltage antinodes. Great care must then be taken by maintenance staff who are required to work on, or near to, the feeder system.

Measurement of V.S.W.R.

The v.s.w.r. on a mismatched transmission line can be determined by measuring the maximum and minimum voltages that are present on the line. In practice, the measurement is generally carried out using an instrument known as a STANDING-WAVE INDICATOR. Measurement of v.s.w.r. not only shows up the presence of reflections on a line but it also offers a most convenient method of determining the nature of the load impedance.

The measurement procedure is as follows. The v.s.w.r. is measured and the distance in wavelengths from the load to the voltage minimum nearest to it is determined. The values obtained allow the magnitude and angle of the voltage reflection coefficient to be calculated. Then, using equation (4.11), the unknown load impedance can be worked out. Unfortunately the arithmetic involved in the latter calculation is fairly lengthy and tedious, and it is customary to use a graphical aid known as the Smith chart which simplifies the work.

If the load impedance is *purely* resistive, a much easier method of measurement is available. Suppose, for example, that $Z_L = R_L = 3R_0$. (Remember that Z_0 is always purely resistive at radio frequencies.) Then,

$$\rho_v = \frac{Z_L - Z_0}{Z_L + Z_0} = \frac{3R_0 - R_0}{3R_0 + R_0} = \frac{1}{2} \underline{/0^\circ}$$

Therefore

$$S = \frac{1 + |\rho_v|}{1 - |\rho_v|} = \frac{1 + \frac{1}{2}}{1 - \frac{1}{2}} = 3$$

Now suppose that instead $Z_L = R_L = \frac{1}{3}R_0$. Then

$$\rho_v = \frac{\frac{1}{3}R_0 - R_0}{\frac{1}{3}R_0 + R_0} = \frac{1}{2} \underline{/180^\circ}$$

and

$$S = \frac{1 + \frac{1}{2}}{1 - \frac{1}{2}} = 3 \qquad \text{as before}$$

It should be noted that the v.s.w.r. is equal to the ratio R_L/Z_0 or Z_0/R_L. This simple relationship is always true provided the line losses are negligibly small and the load impedance is purely resistive.

EXAMPLE 4.5

The v.s.w.r. on a loss-free line of 50 Ω characteristic impedance is 4.2. Determine the value of the purely resistive load impedance which is known to be larger than 50 Ω.

Solution

$$R_L/Z_0 = S = 4.2$$

Therefore,

$$R_L = SZ_0 = 4.2 \times 50 = 210 \ \Omega \qquad (Ans.)$$

Radio Station Feeders

A feeder is employed to connect an aerial to the radio transmitter or receiver with which it is associated. Feeders are usually either two-wire or twin conductors, about 0.3 m apart mounted on top of wooden poles, or are coaxial pairs. The twin feeders are normally operated *balanced* with respect to earth, but the coaxial feeders have their outer conductor earthed and are therefore *unbalanced.* The two types of feeder have various advantages and disadvantages relative to one another and these are listed below.

(a) Twin feeders are cheaper to provide than coaxial pairs.

(b) It is easier to carry out v.s.w.r. measurements and locate faults on a twin feeder.

(c) Matching devices are commonly used in feeder systems to convert a mismatched aerial load into a more or less matched load. These matching devices are easier and so cheaper to make when manufactured for use in a twin feeder system.

(d) Coaxial feeder is less demanding in its use of space than is twin feeder.

(e) Both the conductors comprising a twin feeder are exposed to the atmosphere and this leads to the two-wire feeder's transmission characteristics being much more variable than those of a coaxial feeder.

(f) At higher frequencies the two conductors of a twin feeder are electrically spaced well apart and tend to radiate energy. Because of this the two-wire feeder has greater losses.

(g) In many radio stations it is often necessary to switch feeders between aerials and between station equipment as propagation conditions vary. The necessary switching arrangements are much easier to augment in conjunction with coaxial feeder than with twin feeder.

Most high-frequency transmitting stations use a combination of the two types of feeder in an attempt to make use of the advantages of each type. Within the radio station building coaxial feeders are used but to connect the building to the aerials the twin feeder is employed.

Exercises

4.1. (a) Why is it necessary to match an aerial to its feeder? (b) What do you understand by the terms (i) reflection coefficient and (ii) voltage standing wave ratio as used with reference to an aerial and feeder system, and what is the relationship between them? (part C & G)

4.2. (a) Why is it necessary to match an aerial to its feeder? (b) What do you understand by the term voltage standing wave ratio? (c) Given a choice between typical balanced and unbalanced feeder systems, which would you choose (i) on a cost basis, (ii) for a high-power handling application, (iii) for a very high frequency application, (iv) to connect a transmitter to a rhombic aerial, (v) to connect a receiver to a Yagi aerial? (part C & G)

4.3. Describe the circumstances in which a standing wave can arise on a transmission line and state the meaning of standing wave ratio. An h.f. transmission line of negligible loss has a characteristic impedance of 600 Ω and is terminated by an aerial. Calculate the standing wave ratio along the line when the aerial impedance is 500 Ω. (part C & G)

4.4. A radio-frequency transmission line has a characteristic impedance of 75 Ω and a phase velocity of propagation of 2.4×10^8 m/s. Determine its inductance and capacitance per metre.

4.5. (a) What is meant by the following terms when applied to a radio-frequency transmission line: (i) voltage reflection coefficient, (ii) voltage standing wave ratio, (iii) wavelength? (b) The v.s.w.r. on a line is 1.05. Calculate its voltage reflection coefficient.

4.6. Describe, in your own words, how a line discontinuity or mismatch will produce standing waves of current and voltage.

4.7. Using the expression for the voltage reflection coefficient of a mismatched transmission line explain why there are no reflections on a correctly terminated line. List the reasons why the presence of a standing wave on an aerial feeder is undesirable.

4.8. A correctly-terminated line is 2000 m in length and has a characteristic impedance of 600 Ω. (a) What is the impedance (i) at the load, (ii) 1000 m from the load, (iii) 1500 m from the load? (b) Explain why the current 100 m from the sending end of the line will lag the sending-end current. (c) The line has an attenuation coefficient of 2 dB/km at a particular frequency. What is its overall loss at twice this frequency?

Short Exercises

4.9. A transmission line is 20 metres long. Is it an electrically short or long line? Give reasons for your answer.

4.10. What is meant by the term characteristic impedance when applied to a radio-frequency transmission line? A line has a characteristic impedance of 60 Ω and its output terminals are connected to the input terminals of a 100 m length of another line. The second line has a characteristic impedance of 60 Ω and is correctly terminated. Determine the input impedance of the first line.

4.11. The v.s.w.r. on a loss-free r.f. line is 5. If the minimum voltage on the line is 1 volt what is the maximum voltage?

4.12. An r.f. line has v.s.w.r. of 2.2. Calculate the magnitude of its current reflection coefficient.

4.13. List the reasons why a high standing wave ratio is undesirable on an aerial feeder.

4.14. What are the minimum and maximum value of v.s.w.r. that can exist on a transmission line? What load impedances do they correspond to?

5 Aerials

Introduction

In a radio communication system, the baseband signal is positioned in a particular part of the frequency spectrum using some form of modulation. The modulated wave is then radiated into the atmosphere, in the form of an electromagnetic wave, by a *transmitting aerial*. For the signal to be received at a distant point, the electromagnetic wave must be intercepted by a *receiving aerial*. The basic principles of aerials, and the meanings of various terms used with them, have been discussed in a previous volume [RSII] and a knowledge of these will be assumed in this chapter. A large number of different kinds of aerial are in existence but only four of them are commonly used in modern radio-telephony systems. These four aerials, namely the Yagi, the rhombic, the log-periodic, and the parabolic reflector, will each be described.

The characteristics of any aerial, such as its efficiency and its radiation pattern are the same whether the aerial is used for transmission or reception, but in this chapter the description will be in terms of transmission. The main differences between practical transmitting and receiving aerials are the (often) tremendously different powers which have to be handled. For example, a transmitting aerial may radiate many kilowatts of power while a receiving aerial may have only a few milliwatts dissipated in it. Other than this, the main requirement of a transmitting aerial is that it should match its feeder in order to ensure maximum power input to the aerial. For a receiving aerial, the priority is for maximum gain and directivity and for minimum sidelobes.

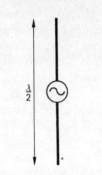

Fig. 5.1 The λ/2 dipole

The Yagi Aerial

The Yagi aerial is made up of a λ/2 dipole and a number of parasitic elements. A λ/2 dipole is a conductor whose electrical length is one-half the wavelength at the desired frequency of operation, and which is centre fed (see Fig. 5.1). The current and voltage distributions in a λ/2 dipole can be deduced from the current and voltage distributions of a λ/4 length of low-loss open-circuited transmission line.

Refer back to Fig. 4.3 which shows the current and voltage standing waves on a λ loss-free line. Over the first λ/4 distance from the open-circuit, the current rises from zero to a maximum, and the voltage falls from a maximum to zero, with the waveforms shown in Fig. 5.2. Similar standing-wave patterns

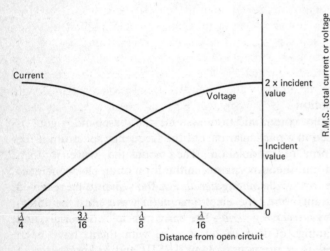

Fig. 5.2 Current and voltage distributions on an open-circuited λ/4 loss-free line

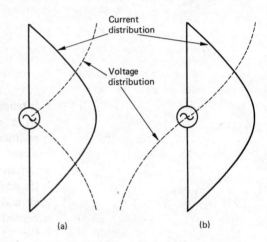

Fig. 5.3 Current and voltage distributions on a λ/2 dipole showing (a) r.m.s. values and (b) peak values

will be obtained if the two conductors forming the line are opened out through an angle of 90° to form a λ/2 dipole aerial. Fig. 5.3a shows the r.m.s. current and voltage distributions on a λ/2 dipole, while Fig. 5.3b shows the distributions when peak values are considered. However, once the conductors are opened out and their separation increased, they will radiate energy. The resultant line losses modify the standing-wave pattern slightly so that the voltage no longer falls to zero at the centre of the dipole.

Impedance of a Dipole

Impedance is the ratio of voltage to current and it is evident from Fig. 5.3a that this ratio is not a constant quantity. At the two ends of the dipole, the voltage is large and the current is small, so the impedance is high, typically about 3500 Ω. At the centre of the dipole the current is large and the voltage is small (not zero because of losses), and the dipole impedance is 73 Ω. This is also the value of the radiation resistance of the aerial. When the impedance of an aerial is referred to, it is necessary to specify the point on the aerial which is to be considered. Usually, as would be expected, the input terminals are chosen.

The INPUT IMPEDANCE of a dipole aerial varies with frequency in the same way as the impedance of a series-tuned circuit. When the aerial is resonant, its electrical length is $\lambda/2$ and its input impedance is purely resistive and equal to 73 Ω. (The physical length is about 5% shorter than $\lambda/2$.) At frequencies higher than the resonant frequency, the signal wavelength is reduced and the dipole becomes electrically longer than $\lambda/2$. Its input impedance is now inductive. When the frequency is reduced below its resonant value, the electrical length of the dipole becomes less than $\lambda/2$ and its input impedance is capacitive. The current fed into the aerial will have its maximum value when the aerial is of resonant length and, since the radiated power is proportional to the square of the current, the aerial is then at its most efficient.

The reactive component of the input impedance of the dipole is a function of the diameter of the conductor. The thicker the conductor the smaller the change in input reactance for a given change in frequency. Thus, when a wide bandwidth is required, a thick conductor must be employed.

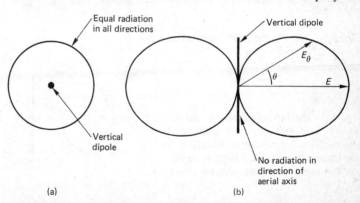

Fig. 5.4 Radiation patterns of a vertical $\lambda/2$ dipole: (a) horizontal plane pattern, (b) vertical plane pattern

The radiation pattern in the horizontal plane of a vertical $\lambda/2$ dipole is a circle as shown in Fig. 5.4a. The RADIATION PATTERN is a graphical representation of the way in which the electric field strength produced by an aerial varies at a fixed distance from the aerial, in all directions in that plane. Hence, the circular radiation pattern means that the dipole will radiate energy equally well in all directions in the horizontal plane. In the vertical plane, the vertical λ/2 dipole does *not* radiate energy equally well in all directions. Indeed, in some directions it does not radiate at all as shown by the radiation pattern of Fig. 5.4*b*. For many radiocommunication systems, other than (most) sound or television broadcasting services, the radiated energy should be concentrated in one or more particular directions, and so some degree of DIRECTIVITY is needed.

An increase in the directivity of a λ/2 dipole can be obtained by the addition of a parasitic element known as a REFLECTOR. A reflector is a conducting rod, approximately 5% longer than λ/2, mounted on the side of the aerial remote from the direction in which maximum radiation should be directed (Fig. 5.5*a*). The reflector is said to be a PARASITIC ELEMENT because it is not electrically connected to the dipole or

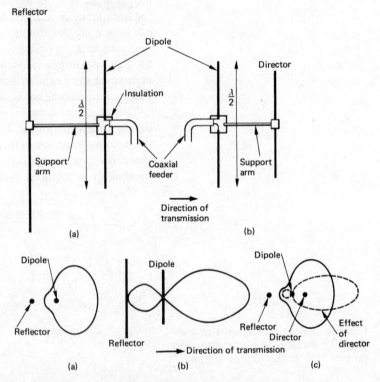

Fig. 5.5 λ/2 dipole with (*a*) a reflector and (*b*) a director

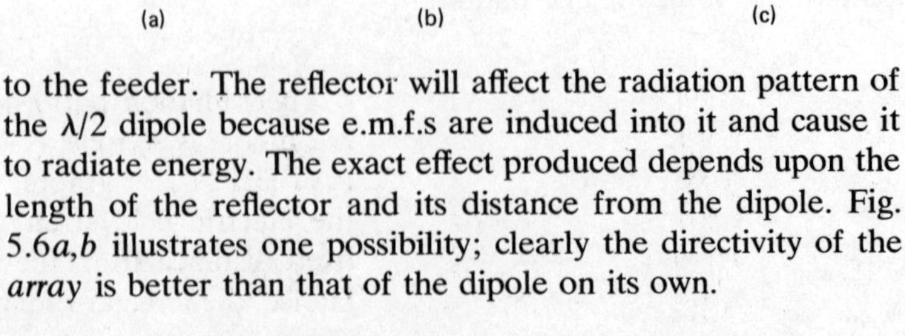

Fig. 5.6 Radiation patterns: (a) λ/2 dipole and reflector: in equatorial plane, (b) λ/2 dipole and reflector: in meridian plane, (c) λ/2 dipole, reflector and director: in equatorial plane

to the feeder. The reflector will affect the radiation pattern of the λ/2 dipole because e.m.f.s are induced into it and cause it to radiate energy. The exact effect produced depends upon the length of the reflector and its distance from the dipole. Fig. 5.6*a,b* illustrates one possibility; clearly the directivity of the *array* is better than that of the dipole on its own.

The action of the reflector effects this improvement as follows. When a voltage, at the resonant frequency of the dipole, is applied to the aerial, an in-phase current flows and the dipole radiates an electromagnetic wave that is in phase with the current. This energy is radiated equally well in all directions perpendicular to the dipole. Some of this energy will arrive at the reflector and induce an e.m.f. in it that will *lag* the voltage applied to the dipole by an angle determined by the element spacing. If, for example, the spacing is 0.15 λ, the induced e.m.f. lags the dipole voltage by 180°. The induced e.m.f. will cause a *lagging* current to flow in the reflector. The reflector will now also radiate energy in all directions normal to it. If *both* the length of the reflector and the dipole/reflector spacing have been chosen correctly, the energy radiated by the reflector will *add* to the energy radiated by the dipole in the wanted direction. Conversely in the opposite direction, i.e. dipole to reflector, the dipole and reflector radiations will subtract from one another.

Further increase in the directivity and gain of a dipole aerial can be achieved by the addition of another parasitic element on the other side of the dipole. This element, known as a DIRECTOR, is made about 5% shorter than the λ/2 dipole. When the dipole radiates energy, an e.m.f. is induced into the director (as well as the reflector) and a *leading* current flows in it. The director then radiates energy in all directions normal to it. The length of the director and its distance from the dipole are both carefully chosen to ensure that the field produced by the dipole is aided in the wanted direction and is opposed in the opposite unwanted direction. The effect of the director on the radiation pattern of the dipole/reflector array can be seen in Fig. 5.6c.

A further increase in the gain and directivity of the aerial cannot be obtained by using a second reflector because the magnetic field behind the reflector has been reduced to small value. The addition of further directors will give extra gain, although the increase per director falls as the number of directors is increased. This is shown by the graph given in Fig. 5.7.

In practice, the choice of element spacing must be a compromise dictated by the gain and front-to-back ratio requirements of the array. Usually, the dipole/reflector spacing is between 0.15 λ to 0.25 λ while the common dipole/director spacing is selected as a value somewhere in the range 0.1 λ to 0.15 λ.

EXAMPLE 5.1

An aerial array consists of a vertical λ/2 dipole with a reflector and

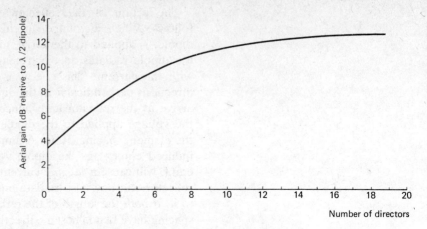

Fig. 5.7 Showing the relationship between the gain of a Yagi aerial and the number of directors employed

one director. Calculate approximate dimensions and spacings for the elements if operation is to be at 100 MHz.

Solution
At 100 MHz $\lambda = 3 \times 10^8 / 100 \times 10^6 = 3$ m
Therefore,

$$\lambda/2 = 1.5 \text{ m}$$

In practice, the dipole would be made slightly shorter because the electric field fringes out at each end of the dipole making its electrical length effectively longer. Therefore,

Dipole length = 1.43 m   (*Ans.*)

The reflector should be about 5% longer than $\lambda/2$ and should be, say, 0.15 λ behind the dipole. Therefore,

Reflector length = 1.57 m   (*Ans.*)

and

Reflector/dipole spacing = 0.6 m   (*Ans.*)

The director should be about 5% shorter than $\lambda/2$. Therefore,

Director length = 1.43 m   (*Ans.*)

and

Director/dipole spacing = 0.4 m   (*Ans.*)

Folded Dipole

The input impedance of a resonant $\lambda/2$ dipole is 73 Ω resistive. The addition of one or more parasitic elements reduces the input impedance to, perhaps, 50 Ω with just a reflector and single director assembly, or perhaps only 20 Ω if several directors are fitted. Normally a 50 Ω or 75 Ω coaxial feeder is used with a Yagi array, and so an impedance mismatch may exist at the aerial terminals which, as shown in Chapter 4, will produce a standing wave pattern on the feeder.

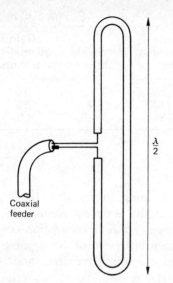

Fig. 5.8 The folded λ/2 dipole

The difficulty could be overcome if the input impedance of the dipole could be increased in some way to a higher value. Then, the reduction in input impedance caused by the addition of parasitic elements would result in an impedance somewhere in the region of the 50 Ω or 75 Ω characteristic impedance of the cable. The higher dipole impedance needed is easily obtained by using a *folded dipole* (see Fig. 5.8.). The input impedance of the folded dipole is *four* times larger than that of the straight dipole, i.e. it is equal to $4 \times 73 = 292$ Ω. Impedance multiplying factors other than four are possible by making the two halves of the folded dipole from different-diameter rods. The bandwidth of the Yagi array is also increased by the use of a folded dipole. A typical Yagi array is shown in Fig. 5.9.

Since the current operation of a Yagi aerial depends critically upon the lengths and spacings of the elements in terms of the signal wavelength, the aerial is only employed at v.h.f. and u.h.f. The physical dimensions necessary to operate in the h.f. band, or even more so in the medium waveband, would make the mechanical structure inconveniently large and correspondingly expensive. The bandwidth of the aerial is the range of frequencies over which the main lobe of its radiation pattern is within specified limits, generally −3 dB, and is of the order of ±3%.

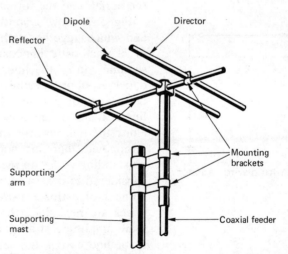

Fig. 5.9 A practical Yagi aerial

Yagi aerials are commonly used for the reception of television broadcast signals in the home and are visible on many rooftops. The aerial also finds considerable application in v.h.f. point-to-point radio-telephony systems for both transmission and reception.

Typical performance figures for Yagi aerials are given in Table 5.1.

Table 5.1

Aerial	Input impedance Ω	Frequency MHz	Bandwidth MHz	Gain dB relative to λ/2 dipole
A	50	90	2	10
B	50	87–100	1–2	10
C	50	47–54	1	8.5

The Rhombic Aerial

Point-to-point radio links operating in the h.f. band (3–30 MHz) are allocated three to five different frequencies to ensure a satisfactory service as sky-wave propagation conditions vary. The h.f. radio transmitters must be capable of rapidly changing frequency as and when required, and it is desirable, for economic reasons, to use the same aerial as much as possible. This requirement rules out the use of a resonant aerial such as the Yagi. A wideband aerial which is used for many h.f. links is the rhombic aerial; the rhombic is a TRAVELLING WAVE TYPE of aerial since its operation depends upon r.f. currents propagating along the full length of the aerial, and the formation of standing waves is avoided.

Fig. 5.10 shows a conductor that is several wavelengths long and which, together with the earth, forms a transmission line of characteristic impedance Z_0 and negligible loss. At the sending-end of the line, a generator of e.m.f. E_s and impedance Z_0 ohms is connected, and at the far end a terminating impedance of Z_0 ohms is used. The input impedance of the line is Z_0 and an input current $I = E_s/2Z_0$ flows and propagates along the line towards the far end. Since the line is correctly terminated, there are no reflections at the load and therefore no standing waves on the line. The line length l metres can be considered to consist of the tandem connection of a very large number of extremely small lengths δl of line. Since the line losses are negligible, the current flowing in the line has the same amplitude at all points. This means that each elemental δl of line carries the same current I and is said to form a CURRENT ELEMENT $I\delta l$.

Each current element will radiate energy. The radiation from each element has its maximum value in the direction making an angle of 90° to the conductor and is zero along the axis of the conductor. The total field strength produced by the line at any point around it is the phasor sum of the field strengths produced by all of the current elements. The phase of the line current varies along the length of the line and in half-wavelength distances a 180° phase shift occurs. Because

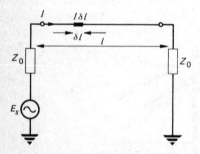

Fig. 5.10 A long-wire radiator

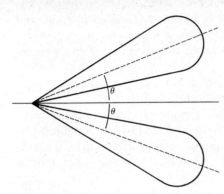

Fig. 5.11 Radiation pattern of a long-wire radiator

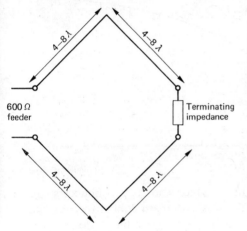

Fig. 5.12 The rhombic aerial

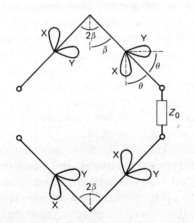

Fig. 5.13 Illustrating the correct choice of tilt angle for a rhombic aerial

of this the field strengths produced by the current elements cancel out in the direction normal to the conductor to produce a null in the radiation pattern. Also, since no element radiates along the axis of the conductor, another null exists in this direction. Thus, the radiation pattern of an electrically long wire is as shown in Fig. 5.11 (a number of small lobes are also present but are not shown). The angle the two main lobes make with the conductor axis is dependent upon the electrical length of the conductor, decreasing as the conductor length is increased. When the length is between 4λ and 8λ, the relationship between lobe angle and frequency is a linear one between limits of $24°$ and $17°$.

The rhombic aerial consists of four such long wires connected together to form, in the horizontal plane, the geometric shape of a rhombus (Fig. 5.12). All the four wires will radiate energy in the directions indicated by its radiation pattern. The aerial is designed so that the main lobes of the radiation pattern of the wires are additive in the wanted direction and self-cancelling in the unwanted directions. This feature of the rhombic aerial is achieved by suitably choosing the angle 2β subtended by two conductors. The TILT ANGLE β (Fig. 5.13) is chosen so that the lobe angle θ is equal to $(90 - \beta)°$. Then the lobes marked X will point in opposite directions and the radiations they represent will cancel, and the lobes marked Y will point in the same (wanted) direction and their radiations will be additive. Since the lobe angle θ varies with frequency, it is not possible to choose the tilt angle to be correct at all the possible operating frequencies, and it is usual to design for optimum operation at the geometric mean of the required frequency band.

EXAMPLE 5.2

A rhombic aerial is to operate over the frequency band 7–14 MHz. Determine a suitable value for the tilt angle.

Solution
The physical lengths of the four wires will be such that at 7 MHz they are 4λ and at 14 MHz they are 8λ long. Then at 7 MHz the lobe angle is $24°$ and at 14 MHz it is $17°$. Since there is a linear relationship between the lobe angle and the frequency of operation, at the geometric mean of 7 MHz and 14 MHz, i.e. at 9.9 MHz, the lobe angle is approximately $20°$. Therefore,

Tilt angle $= 90° - \theta° = 90° - 20° = 70°$ (*Ans.*)

At the design frequency the horizontal plane RADIATION PATTERN of a rhombic aerial is as shown in Fig. 5.14 with the radiated electromagnetic wave being horizontally polarized. The parts of the unwanted lobes that do not cancel are responsible for the sidelobes shown. At frequencies within the

Fig. 5.14 Radiation pattern of a rhombic aerial

bandwidth of the aerial but not at the design frequency, the wanted lobes do not point in exactly the same direction, and the effect on the radiation pattern is to increase its beamwidth and lower its gain in the wanted direction. Typically, a rhombic aerial will operate over a 2 : 1 frequency ratio, e.g. 7–14 MHz with a gain, relative to a λ/2 dipole, that varies with frequency in the manner indicated by Fig. 5.15.

Since the rhombic aerial is primarily used with sky-wave propagation systems, the energy radiated by the aerial must be directed towards the ionosphere at the correct angle of elevation. The ANGLE OF ELEVATION of the main lobe is

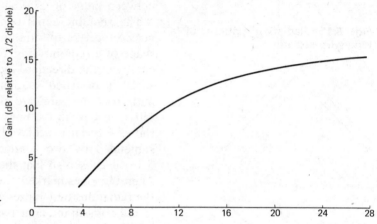

Fig. 5.15 Gain/frequency characteristic of a rhombic aerial

determined by the height at which the four conductors are mounted above the earth. For the main lobe to be at the required angle of elevation, the energy radiated downwards towards the ground must be reflected at such an angle that it is then in *phase* with the directly-radiated energy (Fig. 5.16). The field strength produced in the wanted direction is then doubled, which corresponds to an increase in gain of 6 dB.

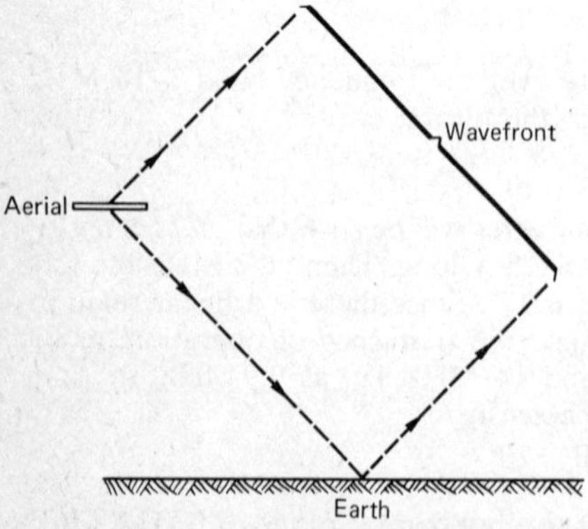

The INPUT IMPEDANCE of a rhombic aerial is determined by both the signal frequency and the diameter of the wires and is in the range 600 Ω to 800 Ω. The input impedance is frequency-dependent and can be made more nearly constant by using more than one wire to form each arm of the rhombus. The input impedance is then also a function of the number of wires used and their distance apart.

The TERMINATING IMPEDANCE must match the characteristic impedance of the lines forming the aerial in order to prevent reflections taking place and this means that one-half of the power fed into the aerial will be dissipated in the terminating impedance; the aerial, therefore, has a maximum efficiency of 50%. When the powers involved are small, as with a receiving rhombic aerial, a carbon resistor will often suffice as the terminating impedance. For high-power installa-

Fig. 5.16 Optimum height above ground for a rhombic aerial

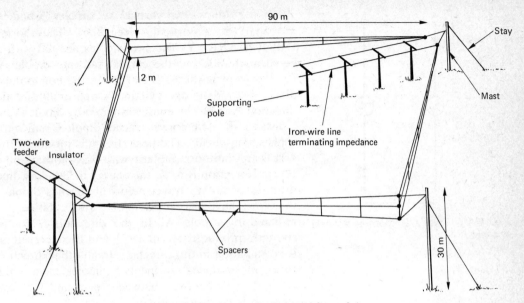

Fig. 5.17 A practical rhombic aerial

tions, such a simple solution is not available and commonly the impedance consists of a two-wire line using iron conductors.

A typical rhombic aerial is illustrated by Fig. 5.17.

The rhombic aerial possesses two disadvantages which are tending to lead to its replacement by the log-periodic aerial. Firstly, its radiation pattern exhibits relatively large sidelobes. Sidelobes in a radiation pattern are undesirable since, in a transmitting aerial, they mean that power is radiated in unwanted directions. The energy may interfere with other systems but, in any case, represents a waste of power. In the case of a receiving aerial the unwanted sidelobes indicate a response to interference and noise arriving from unwanted directions. The second disadvantage, made clear by the typical dimensions given in Fig. 5.17, is the large site area which must be provided to accommodate a rhombic aerial.

The Log-Periodic Aerial

The log-periodic aerial provides an alternative to the rhombic aerial in the h.f. band and is particularly good when the available site area is limited and/or an elevation angle in excess of about 40° is required. The aerial can operate over a wide frequency band and has very small side- and back-lobes.

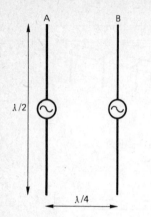

A　　　　　　B

λ/2

λ/4

Fig. 5.18 Two λ/2 dipoles spaced λ/4 apart

Fig. 5.18 shows two vertical λ/2 dipoles A and B which are λ/4 apart at a particular frequency. They form a TWO-DIPOLE ARRAY. The two dipoles are fed with currents of equal amplitude but the current fed into dipole A leads the input current of dipole B by 90°. In the horizontal plane each dipole will radiate energy equally well in all directions. In the direction A to B, the energy radiated by dipole A has to travel a distance of λ/4 before it reaches dipole B and will experience a phase lag of 90°. The field strengths produced by dipoles A and B are therefore in phase with one another and add. In the reverse direction B to A, the energy radiated by dipole B has a distance of λ/4 to travel before it reaches dipole A. It will, therefore, have a total phase of −180° relative to the energy radiated by dipole A. In the direction B to A the field strengths produced by dipoles A and B cancel out and so there is no radiation in this direction. In all other directions the field strengths produced by the two dipoles have a phase angle, other than 0° or 180°, between them and the resultant field strength is given by their phasor sum.

The RADIATION PATTERN of the two-dipole array is shown in Fig. 5.19. Greater directivity and gain can be achieved by the addition of a third dipole C, λ/4 apart from dipole B. Dipole C is fed with a current of equal amplitude to, but lagging by 90°, the current into dipole B. Similarly, a fourth, a fifth, and more dipoles can be added to further increase the directivity of the aerial. Unfortunately, the operation of an *end-fire array* depends critically on the spacings of its component dipoles, and this means that the aerial is only suitable for operation at a single frequency.

The log-periodic aerial provides an end-fire radiation pattern over a wide frequency band. The aerial is manufactured in various forms and Fig. 5.20a shows an example of the log-periodic dipole array which is used at v.h.f. The lengths l_1, l_2, l_3, etc. of the dipoles increase from left to right with the relationship $l_2/l_1 = l_3/l_2 = l_4/l_3$ etc., the common ratio being known as the *scale factor* τ of the aerial. The spacings d_1, d_2, d_3, etc. between adjacent dipoles also increase from left to right. Successive spacings are also related by the same scale factor τ.

The input signal is applied to the aerial via a twin feeder and is applied to adjacent dipoles with 180° phase change because the connections to successive dipoles are reversed. At any given frequency within the bandwidth of the aerial, only two, or, perhaps, three dipoles are at or anywhere near resonance, i.e. approximately λ/2 long. These dipoles take a relatively large input current and radiate considerable energy; because of the phasing of the dipole currents, an end-fire effect is

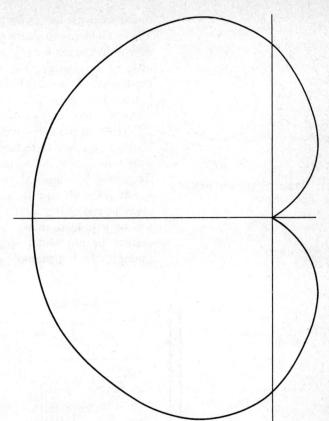

Fig. 5.19 Horizontal plane radiation pattern of two vertical λ/2 dipoles spaced λ/4 apart, fed with equal amplitude currents

Fig. 5.20 (*a*) The log-periodic dipole aerial: general principle

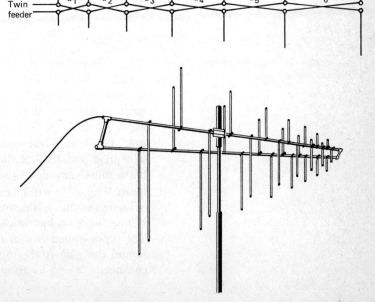

Fig. 5.20 (*b*) Practical example of a log-periodic aerial

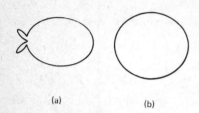

(a)   (b)

Fig. 5.21 Radiation pattern of a log-periodic dipole aerial

obtained, with the main lobe being in the direction of from longer elements to shorter elements. All the other dipoles are now either much longer or much shorter than $\lambda/2$ and radiate little or no energy. The radiation pattern of a log-periodic dipole aerial is shown in Fig. 5.21. The equatorial plane pattern is given in *a* and the meridian plane pattern in *b*.

As the frequency of the input signal is varied, the ACTIVE REGION of the aerial will move in one direction or the other. If the frequency is reduced, the active region will move towards the end of the aerial where the dipoles are longer. If the frequency is increased, shorter dipoles become resonant or nearly resonant and the active region moves towards the short element end of the aerial. The dipole array can be mounted on top of a pole or mast and oriented to operate with either vertical or horizontal elements. Fig. 5.20*b* shows a typical example of a log-periodic dipole aerial.

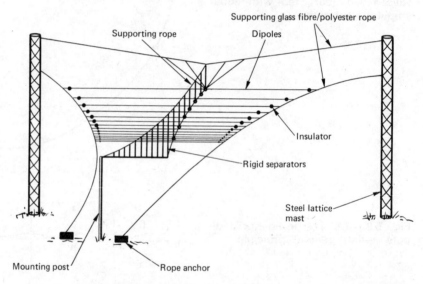

Fig. 5.22 A high-frequency log-periodic aerial

A HIGH-FREQUENCY VERSION of the log-periodic aerial is shown in Fig. 5.22. Since the aerial is mounted close to the earth, its vertical plane radiation pattern is modified by earth reflections. If it is desired to have the same elevation angle for the radiation pattern at all frequencies, each element must be at the same *electrical height* above the ground. This, of course, means that the physical height of the aerial above earth must vary along the length of the aerial as shown in the figure.

Log-periodic aerials are used for h.f. communication links where the wide bandwidth, 4–30 MHz, is often an advantage. The input impedance of the aerial lies in the range of 50–300 Ω and the gain is about 10 dB relative to a $\lambda/2$ dipole, which can be increased to about 14 dB by earth reflections.

The gain is less than that of the rhombic, which is an indication of a radiation pattern that is not very directive. On the other hand its side- and back-lobe levels are small.

Compared to the Yagi aerial, the v.h.f. log-periodic aerial has a smaller gain (for the same number of elements), and smaller side- and back-lobes.

The Parabolic Reflector Aerial

At frequencies at the upper-end of the u.h.f. band and in the s.h.f. band, the signal wavelength becomes sufficiently small to allow a completely different kind of aerial to be used. The aerial is known as the parabolic reflector or DISH aerial and it is capable of producing a very directive, high-gain radiation pattern.

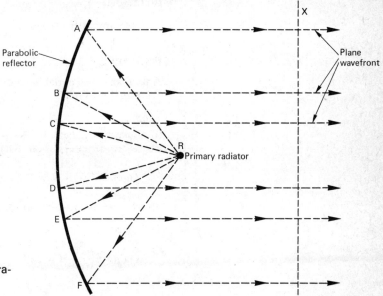

Fig. 5.23 Reflection from a parabolic reflector

The aerial consists of a large metal dish which is used to reflect into the atmosphere the radio energy directed onto it by a smaller radiator (often a dipole/reflector array) mounted at the focal point. The idea of the aerial is illustrated by Fig. 5.23. A property of a parabolic dish is that the distance from the focal point of the dish to an arbitrary plane the other side of the focal point is a constant regardless of which point on the dish is considered, i.e.

$$RAX = RBX = RCX = RDX = REX = RFX$$

Because of this property, the spherical wavefront signal originating from the radiator and reflected by the dish arrives at the plane X with a plane wavefront. The reflected waves are all parallel to one another and form a concentrated, highly-directive radio wave.

When used as a receiving aerial, the action of the dish is reversed; the incoming plane wavefront radio wave is reflected by the dish and brought to a focus at the focal point where a small receiving aerial, such as a $\lambda/2$ dipole, is mounted and is connected to the feeder.

The GAIN of a parabolic dish aerial depends upon its diameter in terms of the signal wavelength. If the diameter is made several times larger than the signal wavelength, very high gains can be obtained. The relationship between aerial gain and the diameter D of the dish is given by equation (5.1):

$$\text{Gain} = 6(D/\lambda)^2 \tag{5.1}$$

The beamwidth of the aerial is also a function of the dish diameter;

$$\text{Beamwidth} = 70\ \lambda/D \tag{5.2}$$

EXAMPLE 5.3

A parabolic dish aerial has a diameter of 1 m. Determine its gain at (i) 1 GHz, (ii) 6 GHz.

Solution
(i) Signal wavelength $\lambda = 3 \times 10^8/10^9 = 0.3$ m
Therefore, from equation (5.1)

$$\text{Gain} = 6\left(\frac{1}{0.3}\right)^2 = 66.7 \quad (Ans.)$$

(ii) $\lambda = 3 \times 10^8/6 \times 10^9 = 0.05$ m

Therefore,

$$\text{Gain} = 6\left(\frac{1}{0.05}\right)^2 = 2400 \quad (Ans.)$$

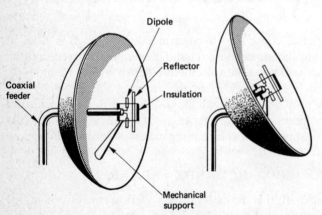

Fig. 5.24 Two views of a parabolic dish aerial

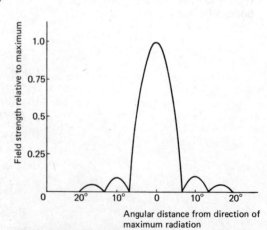

Fig. 5.25 Radiation pattern of a parabolic dish aerial

Fig. 5.24 shows the appearance of a practical dish aerial.

The radiation pattern of a parabolic reflector has one main, *very*-narrow-beamwidth lobe and a number of much smaller sidelobes. The main lobe is so narrow that the radiation pattern cannot conveniently be plotted in the usual manner. Usually, the radiation pattern is only drawn for a small angular distance either side of the direction of maximum radiation, and Fig. 5.25 shows a typical pattern.

Exercises

5.1. From sketches of the radiation characteristics of a single non-resonant wire develop the radiation patterns showing the directional characteristics of a rhombic aerial. How do these characteristics vary with frequency? Describe how a transmitting rhombic aerial is terminated and describe the constructional features which provide a good impedance match over the required frequency band. *(C & G)*

5.2. Describe the construction and action of a paraboloid aerial and sketch a typical radiation pattern. How are the gain and beamwidth of such an aerial related to its diameter and the frequency of transmission? Sketch the device used for launching the radio energy into the aerials. *(Part C & G)*

5.3. Explain clearly what is meant by the gain of an aerial. Sketch and describe a rhombic aerial suitable for h.f. transmission. What are the main design features that determine its performance? State typical values for (*a*) the gain, (*b*) the frequency bandwidth, and (*c*) the angle of elevation of the main lobe. *(C & G)*

5.4. Sketch and give dimensions of a high-power wide-bandwidth transmitting aerial used for long distance point-to-point h.f. services. Describe the method of feeding the aerial. How does the gain of the aerial vary with frequency? *(C & G)*

5.5. Sketch, and give approximate dimensions in terms of wavelength for, a v.h.f. yagi aerial comprising a folded dipole with a reflector and director. For such an aerial state clearly what is meant by the terms (*a*) gain, (*b*) beamwidth, (*c*) bandwidth. Explain the purpose of the folded dipole. *(C & G)*

5.6. (*a*) What do you understand by the gain of an aerial? (*b*) Draw a dimensioned sketch of a typical yagi aerial. (*c*) Give typical values for such an aerial of (i) gain, (ii) frequency band of operation. (*d*) Why is the yagi not commonly used at m.f.? (*e*) Why is the aerial fed via a folded dipole? *(C & G)*

5.7. (*a*) Draw a dimensioned sketch of a rhombic aerial. (*b*) Why are three wires often employed on each leg of a rhombic aerial? (*c*) Upon what factors does the impedance of a rhombic aerial depend? (*d*) Why is it desirable to be able to alter the height of a rhombic aerial? (*e*) What is the most noticeable difference between a rhombic aerial used for receiving and one used for transmitting?

5.8. (*a*) Draw a dimensioned sketch of a yagi aerial suitable for operation at 200 MHz and explain why a folded dipole is normally used. (*b*) Draw the horizontal and vertical radiation patterns of (i) a vertical dipole, (ii) a vertical dipole with one reflector. (*c*) Show how the gain of an aerial relative to a dipole can be obtained from its radiation pattern. (*C & G*)

5.9. Complete the table of comparison, Table 5.2, for the yagi, rhombic and log-periodic aerials as applied to a typical aerial of each type. (*C & G*)

Table 5.2

	RHOMBIC	LOG-PERIODIC	YAGI
Operating frequency range			
Bandwidth			
Gain			
Input impedance			
Dimensions			
For what service is it commonly used?			
Is it a travelling-wave aerial?			

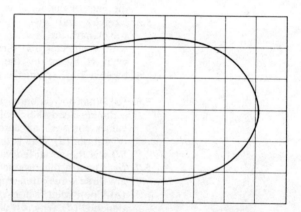

Fig. 5.26

5.10. Fig. 5.26 shows the horizontal radiation pattern of an aerial. Determine the gain of this aerial relative to a vertical $\lambda/2$ dipole.

5.11. Sketch the construction of log-periodic aerial suitable for use in the h.f. band. Give typical values of gain and input impedance and state whether your aerial produces a horizontally- or a vertically-polarized wave.

5.12. Sketch the radiation pattern of a half-wave dipole in both the equatorial and meridian planes. Show the current and voltage distributions in the aerial. Draw the arrangement of a simple yagi aerial with one director and show how the radiation pattern differs from that of the dipole alone. Label your aerial sketch with dimensions suitable for use at a frequency of (a) 45 MHz, (b) 490 MHz.

5.13. When an aerial radiates a power of 1 kW, a field strength of 10 mV/m is set up at a certain distant point. Another aerial located alongside the first needs only to radiate 250 W to produce the same field strength at the same point. If the first aerial has a gain of 20 dB relative to an isotropic radiator, determine the gain of the second aerial.

5.14. Describe the way in which energy is radiated from a conductor carrying a high-frequency current. Hence explain why aerials for use at v.h.f. are more efficient than those used at medium frequencies. Quote typical figures for aerial efficiency. An aerial has a loss resistance of 3 Ω and a radiation resistance of 1 Ω. If the current fed into the aerial is 10 A calculate the aerial efficiency, and the power radiated.

5.15. Explain the meanings of the following terms used in aerial work:
(a) polarization (b) gain (c) radiation pattern
(d) beamwidth (e) isotropic radiation
(f) front-to-back ratio (g) parasitic element

Short Exercises
5.16. Draw the radiation patterns of a horizontal λ/2 dipole in (a) the vertical plane and (b) the horizontal plane.

5.17. Sketch, with dimensions in metres, a λ/2 dipole with a reflector and one director array suitable for operation at 100 MHz.

5.18. A dipole is 1.2 m long. At what frequency is this dipole a half-wavelength long?

5.19. Write down typical figures for the gain and the input impedance of (i) a yagi aerial, (ii) a rhombic aerial, (iii) a log-periodic aerial.

5.20. What is meant by (i) the radiation pattern, (ii) the beamwidth, and (iii) the gain of an aerial?

5.21. Explain why directivity in a receiving aerial is desirable.

5.22. Draw a folded λ/2 dipole and say why and when it is used.

5.23. A parabolic reflector is to have a gain of 1000. Calculate the diameter required if operation is at (i) 3 GHz, (ii) 3 MHz. Comment on your answer.

6 The Propagation of Radio Waves

Introduction

When a radio-frequency current flows into a transmitting aerial, a radio wave at the same frequency is radiated in a number of directions as predicted by the radiation pattern of the aerial. The radiated energy will reach the receiving aerial or, in the case of broadcast or mobile systems, receiving aerials, by one or more of five different modes of propagation. Four of these modes, the *surface wave*, the *sky wave*, the *space wave* and the use of a communication satellite, are illustrated by Fig. 6.1.

The surface wave is supported at its lower edge by the surface of the earth and is able to follow the curvature of the earth as it travels. The sky wave is directed upwards from the earth into the upper atmosphere where, if certain conditions are satisfied, it will be returned to earth at the required locality. The space wave generally has two components, one of which travels in a very nearly straight line between the transmitting and receiving aerials, and the other which travels by means of a single reflection from the earth. The fourth method illustrated is a relatively modern technique that utilizes the ability of a *communication satellite* orbiting the earth to receive a signal, amplify it, and then transmit it at a different frequency towards the earth. The fifth method of propagation which is not shown in Fig. 6.1 is known as *scatter* and is used only when, for one reason or another, one of the other methods is not available.

The radio frequency spectrum has been subdivided into a number of frequency bands and these are given in Table 6.1.

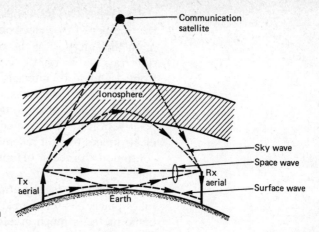

Fig. 6.1 Modes of propagation

Table 6.1

Frequency band	Classification	Abbreviation
10–30 kHz	Very low	v.l.f.
30–300 kHz	Low	l.f.
300–3000 kHz	Medium	m.f.
3–30 MHz	High	h.f.
30–300 MHz	Very high	v.h.f.
300–3000 MHz	Ultra high	u.h.f.
3–30 GHz	Super high	s.h.f.

The surface wave is used for world-wide communications in the v.l.f. and l.f. bands and for broadcasting in the m.f. bands.

The sky wave is used for h.f. communication systems, including long-distance radio-telephony and sound broadcasting.

The space wave is used for sound and television broadcasting, for multi-channel telephony systems, and for various mobile systems operating in the v.h.f., and u.h.f. and s.h.f. bands.

Communication satellite systems are used to carry international multi-channel telephony systems and sometimes television signals.

Lastly, scatter systems operate in the u.h.f. band to provide multi-channel telephony links.

The Ionosphere

Ultra-violet radiation from the sun entering the atmosphere of the earth supplies energy to the gas molecules of the atmosphere. This energy is sufficient to produce ionization of the molecules, that is remove some electrons from their parent atoms. Each atom losing an electron in this way has a resultant positive charge and is said to be *ionized*.

The IONIZATION thus produced is measured in terms of the number of free electrons per cubic metre and is dependent upon the intensity of the ultra-violet radiation. As the radiation travels towards the earth, energy is continually extracted from it and so its intensity is progressively reduced.

The liberated electrons are free to wander at random in the atmosphere and in so doing may well come close enough to a positive ion to be attracted to it. When this happens, the free electron and the ion recombine to form a neutral atom. Thus a continuous process of ionization and recombination takes place.

At high altitudes the atmosphere is rare and little ionization takes place. Nearer the earth the number of gas molecules per cubic metre is much greater and large numbers of atoms are ionized; but the air is still sufficiently rare to keep the probability of recombination at a low figure. Nearer still to the earth, the number of free electrons produced per cubic metre falls, because the intensity of the ultra-violet radiation has been greatly reduced during its passage through the upper atmosphere. Also, since the atmosphere is relatively dense the probability of recombination is fairly high. The density of free electrons is therefore small immediately above the surface of the earth, rises at higher altitudes, and then falls again at still greater heights. The earth is thus surrounded by a wide belt of ionized gases, known as the IONOSPHERE.

In the ionosphere, LAYERS exist within which the free electron density is greater than at heights immediately above or below the layer. Four layers exist in the daytime (the D, E, F_1 and F_2 layers) at the heights shown in Fig. 6.2.

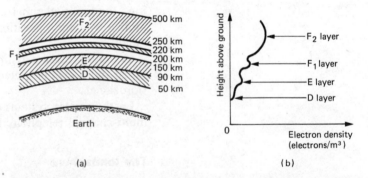

Fig. 6.2 Layers in the ionosphere

The heights of the ionospheric layers are not constant but vary both daily and seasonally as the intensity of the sun's radiation fluctuates. The electron density in the D-layer is small when compared with the other layers. At night-time when the ultra-violet radiation ceases, no more free electrons are produced and the D-layer disappears because of the high rate of recombination at the lower altitudes. The E-layer is at a height of about 100 km and so the rate of recombination is smaller. Because of this, the E-layer, although becoming weaker, does not normally disappear at night-time. In the daytime, the F_1 layer is at a more or less constant height of 200–220 km above ground but the height of the F_2 layer varies considerably. Typical figures for the height of the F_2 layer are 250–350 km in the winter and 300–500 km in the summer.

The region of the earth's atmosphere between the surface of the earth and the lower edge of the ionosphere is known as the TROPOSPHERE. The behaviour of the ionosphere when a radio wave is propagated through it depends very much upon the frequency of the wave. At low frequencies the ionosphere acts as though it were a medium of high electrical conductivity and *reflects*, with little loss, any signals incident on its lower edge. It is possible for a v.l.f. or l.f. signal to propagate for considerable distances by means of reflections from both the lower edge of the ionosphere and the earth. This is shown by Fig. 6.3. The wave suffers little attenuation on each reflection and so the received field strength is inversely proportional to the distance travelled.

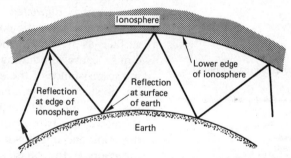

Fig. 6.3 Multi-hop transmission of a low-frequency wave

In the m.f. band the D-layer acts as a very lossy medium whose attenuation reaches its maximum value at a frequency of 1.4 MHz, often known as the *gyro-frequency*. Generally, m.f. signals suffer so much loss in the D-layer that little energy reaches the E or F layers. At night-time, however, the D-layer has disappeared and an m.f. signal will be *refracted* by the E-layer and perhaps also by the F-layer(s) and returned to earth. With further increase in frequency to the h.f. band, the ionospheric attenuation falls and the E and F layers provide *refraction* of the sky wave. At these frequencies the D-layer has little, if any, refractive effect but it does introduce some losses.

The amount of refraction of a radio wave that an ionospheric layer is able to provide is a function of the frequency of the wave, and at v.h.f. and above no useful reflection is obtained (usually). This means that a v.h.f. or s.h.f. signal will normally pass straight through the ionosphere.

The Ground or Surface Wave

At v.l.f. and l.f. the transmitting aerial is electrically short but physically very large and must therefore be mounted vertically on the ground. The aerial will radiate energy in several directions and produce both surface and space waves (sometimes the sky wave too). The combined surface and space wave is known as the GROUND WAVE. At these frequencies the signal wavelength is long and the aerial height is only a small fraction of a wavelength. The reflected component of the space wave experiences a 180° phase shift upon reflection, and since the difference, in wavelengths, between the lengths of the direct and reflected waves is very small, the two waves cancel out. Because of this, v.l.f and l.f propagation is predominantly by means of the surface wave. Very often the term ground wave is used to represent the surface wave.

The surface or ground wave is one which leaves the transmitting aerial very nearly parallel to the ground. Vertically polarized waves must be used because horizontal polarization would result in the low resistance of the earth short-circuiting the electric component of the wave. The surface wave follows the curvature of the earth as it travels from the transmitter because it is *diffracted.†* Further bending of the wave occurs because the magnetic component of the wave cuts the earth's surface as it travels and induces e.m.f.s in it. The induced e.m.f.s cause alternating currents to flow and dissipate power in the resistance of the earth. This power can only be supplied by the surface wave, and so a continuous flow of energy from the wave into the earth takes place. The signal wavefront, therefore, has two components of velocity, one in the forward direction and one downwards towards the earth. The resultant direction is the phasor sum of the forward and downward components, and this results in the wave being tilted forward, as shown in Fig. 6.4. The downward component is always normal to the earth and the forward component 90° advanced; hence the tilted wavefront follows the undulations of the ground (Fig. 6.5).

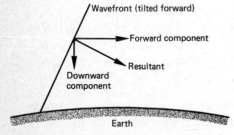

Fig. 6.4 Wavefront of the surface wave.

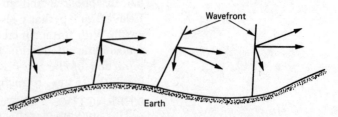

Fig. 6.5 Propagation of surface wave over undulating terrain

† Diffraction is a phenomenon which occurs with all wave motion. It causes a wave to bend round any obstacle it passes. For a surface wave, the earth itself is the obstacle.

The transfer of energy from the wave to the ground attenuates the wave as it travels, and the field strength E_d at a distance d kilometres from the transmitter is given by

$$E_d = K\frac{E_1}{d} \tag{6.1}$$

where E_1 is the field strength 1 km from the transmitter and K is a factor representing the wave attenuation caused by the power dissipated in the ground.

The attenuation factor K depends upon the frequency of the wave, and the conductivity and permittivity of the earth. The attenuation at a given frequency is least for propagation over expanses of water and greatest for propagation over dry ground, such as desert. For propagation over ground of average dampness, with a radiated power of 1 kW, the distance giving a field strength of 1 mV/m varies approximately with frequency as shown in Table 6.2.

Table 6.2

Frequency	Range (km)
100 kHz	200
1 MHz	60
10 MHz	6
100 MHz	1.5

At m.f, particularly at the higher end of the band, the height of the aerial is a much larger fraction of the signal wavelength. Now complete cancellation of the direct and reflected components of the space wave no longer occurs and the space wave partially contributes to the field strength over shorter distances.

Refraction of an Electromagnetic Wave

When an electromagnetic wave travelling in one medium passes into a different medium, its direction of travel will probably be altered. The wave is said to be *refracted*. The ratio

$$\frac{\text{sine of angle of incidence, } \phi_i}{\text{sine of angle of refraction, } \phi_r}$$

is a constant for a given pair of media and is known as the REFRACTIVE INDEX for the media. If one of the two media is air the *absolute refractive index* of the other medium is obtained.

If a wave passes from one medium to another medium that has a lower absolute refractive index, the wave is bent away from the normal (Fig. 6.6*a*). Conversely, if the wave travels

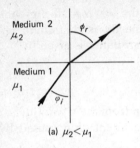

(a) $\mu_2 < \mu_1$

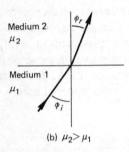

(b) $\mu_2 > \mu_1$

Fig. 6.6 Refraction of electromagnetic waves: wave passing into a medium of (a) lower absolute refractive index; (b) higher absolute refractive index

into a region of higher absolute refractive index, the wave is bent towards the normal (Fig. 6.6b).

Suppose a wave is transmitted through a number of thin parallel strips (Fig. 6.7), each strip having an absolute refractive index lower than that of the strip immediately below it. The wave will pass from higher to lower absolute refractive index each time it crosses the boundary between two strips, and it is therefore progressively bent *away* from the normal. If the widths of the strips are made extremely small, the absolute refractive index will steadily decrease and the wave will be continuously refracted.

Within an ionospheric layer the electron density increases with increase in height above the earth. Above the top of the layer, the density falls with further increase in height until the lower edge of the next, higher layer is reached. At heights greater than the top of the F_2-layer, the electron density falls until it becomes negligibly small.

The *refractive index n* of a layer is related to both the frequency of the wave and the electron density according to equation (6.2):

$$n = \frac{\sin \phi_i}{\sin \phi_r} = \sqrt{\left(1 - \frac{81N}{f^2}\right)} \qquad (6.2)$$

Here f is the frequency of the radio wave in hertz, N is the number of free electrons per cubic metre, and as before ϕ_i and ϕ_r are respectively the angles of incidence and refraction.

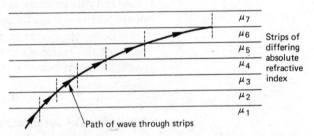

Fig. 6.7 Refraction of an electromagnetic wave passing through media of progressively lower absolute refractive index:

Equation (6.2) shows that the refractive index of a layer decreases as the electron density is increased. This means that *within a layer* the refractive index *falls* with increase in height above ground. Also to be noted is that an increase in frequency results in an increase in the refractive index of a layer.

A radio wave at a particular frequency entering a layer with angle of incidence ϕ_i will always be passing from lower to higher refractive index as it travels upwards through the layer. Therefore, the wave is continuously refracted away from the normal. If, before it reaches the top of the layer, the wave has been refracted to the extent that the angle of refraction ϕ_r becomes equal to 90°, the wave will be returned to earth.

Should the angle of refraction be less than 90°, the wave will emerge from the top of the layer and travel on to a greater height. If, then, the wave enters another, higher layer, it will experience further refraction and may now be returned to earth. If the frequency of the wave is increased, the wave will be refracted to a lesser extent and will have to travel further through a layer before it is returned to earth.

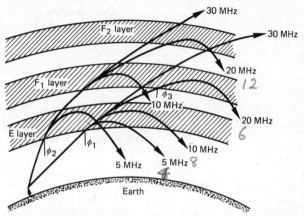

Fig. 6.8 Effect on ionospheric refraction of angle of incidence and frequency of wave

Suppose sky waves at frequencies 5, 10, 20 and 30 MHz are transmitted and are incident on the lower edge of the E-layer with an angle of incidence ϕ_1 (Fig. 6.8). The 5 MHz wave is refracted to the greatest extent and is returned to earth after penetrating only a little way into the E-layer. The 10 MHz wave must penetrate much farther into the E-layer before it is returned to earth, while the 20 MHz wave is hardly refracted at all by the E-layer and passes on to the F_1-layer. The 20 MHz wave meets the F_1-layer with a much larger angle of incidence, $\phi_3 > \phi_1$. A smaller change in direction is now required to return the wave to earth, and sufficient refraction is produced by the F_1-layer. The 30 MHz wave is not refracted to the extent required to return it to earth and escapes from the top of the F_2-layer.

If the angle at which the waves are incident on the E-layer is reduced to ϕ_2, greater refraction is necessary to return the wave to earth. Consequently, only the 5 MHz wave is now returned by the E-layer, the 10 MHz and 20 MHz waves passing right through and arriving at the F_1-layer. The refractive index of the F_1-layer is lower at 10 MHz than at 20 MHz; hence the 10 MHz wave is refracted sufficiently to be returned, but the 20 MHz wave is not. The 20 MHz wave passes on to the F_2-layer and is then returned. Once again the 30 MHz wave is not returned.

Further decrease in the angle of incidence of the waves on the E-layer may well result in the 20 MHz wave escaping the F_2-layer also and not returning to earth at all, the 5 MHz and 10 MHz waves being returned by a higher layer.

EXAMPLE 6.1

An ionospheric layer has a maximum electron density of 6×10^{11} electrons/m^3. Calculate the maximum frequency that will be returned to earth if the angle of incidence is (i) 60°, (ii) 30°.

Solution

(i) From equation (6.2)

$$\sin 60° = 0.866 = \sqrt{\left(1 - \frac{81 \times 6 \times 10^{11}}{f^2}\right)}$$

$$0.75 = 1 - \frac{81 \times 6 \times 10^{11}}{f^2}$$

$$f^2 = \frac{81 \times 6 \times 10^{11}}{0 \cdot 25}$$

$$f = 13.943 \text{ MHz} \qquad (Ans.)$$

(ii) $$\sin 30° = 0.5 = \sqrt{\left(1 - \frac{81 \times 6 \times 10^{11}}{f^2}\right)}$$

$$0.25 = 1 - \frac{81 \times 6 \times 10^{11}}{f^2}$$

$$f^2 = \frac{81 \times 6 \times 10^{11}}{0.75}$$

$$f = 8.05 \text{ MHz} \qquad (Ans.)$$

Critical Frequency

The critical frequency of an ionospheric layer is the *maximum* frequency that can be radiated vertically upwards by a radio transmitter and be returned to earth. This condition corresponds to a wave that travels to the top of the layer, where the electron density is at its maximum value, before its angle of refraction becomes 90°. The angle of incidence is 0°. Therefore, using equation (6.2),

$$\sin 0° = 0 = \sqrt{\left(1 - \frac{81 N_{max}}{f_{crit}^2}\right)}$$

Therefore,

$$f_{crit}^2 = 81 N_{max}$$
$$f_{crit} = 9\sqrt{N_{max}} \qquad (6.3)$$

The critical frequency of a layer is of interest for two reasons: firstly it is a parameter which can be measured from the ground and, secondly, it bears a simple relationship to the *maximum usable frequency* of a sky-wave link.

Maximum Usable Frequency

The maximum usable frequency (m.u.f.) is the highest frequency that can be used to establish communication, using the sky wave, between two points. If a higher frequency is used, the wave will escape from the top of the layer and the signal will not be received at the far end of the link. The m.u.f. is determined by both the angle of incidence of the radio wave and the critical frequency of the layer; thus

$$\text{m.u.f.} = f_{crit}/\cos \phi_i \qquad\qquad (6.4)$$

The m.u.f. is an important parameter in sky-wave propagation. Since the attenuation suffered by a wave is inversely proportional to the frequency of the wave it is desirable to use as high a frequency as possible.

EXAMPLE 6.2

Calculate the maximum usable frequency of a sky-wave link if the angle of incidence is 45° and the maximum electron density of the layer used is 4×10^{11} electrons/m³.

Solution
From equation (6.3)

$$f_{crit} = 9\sqrt{(4 \times 10^{11})} = 5.692 \text{ MHz}$$

Therefore, from equation (6.4)

$$\text{m.u.f.} = 5.692/\cos 45° = 8.05 \text{ MHz} \qquad (Ans.)$$

The electron density of an ionospheric layer is not a constant quantity but is subject to many fluctuations, some regular and predictable and some not. As a consequence the m.u.f. of any given route is also subject to considerable variation over a period of time. The m.u.f. of a link will vary throughout each day as the intensity of the sun's radiation changes. Maximum radiation from the sun occurs at noon, while after dark there is no radiation. There is always a time lag of some hours between a change in the ultra-violet radiation passing through the ionosphere and the resulting change in electron density, and so the m.u.f. may be expected to vary in the manner shown by the typical graphs of Fig. 6.9.

In addition to the predictable m.u.f. variations, further fluctuations often take place and, because of this, operation of a link at the m.u.f. prevailing at a given time would not produce a reliable system. Usually a frequency of about 85% of the m.u.f. is used to operate a sky-wave link. This frequency is known as the *optimum working (traffic) frequency* or o.w.f. Since the m.u.f. will vary over the working day, the o.w.f. will do so also and it is therefore necessary to change the transmitted frequency as propagation conditions vary. The number of

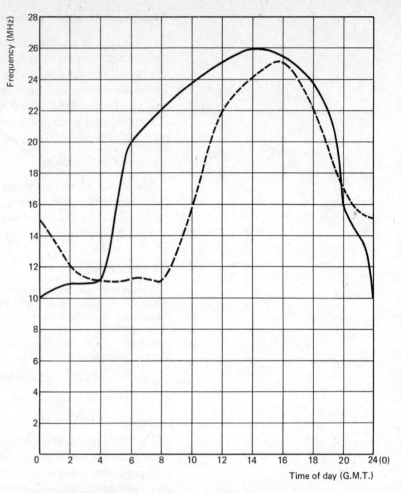

Fig. 6.9 Variations of m.u.f. with time of day

available frequencies is limited and international frequency sharing is necessary. Usually, an individual transmitter is allocated several carrier frequencies, any one of which can be employed if necessary. When propagation conditions are poor, it may prove necessary to transmit on more than one frequency and even, when conditions are particularly bad, to re-transmit when conditions improve.

The attenuation of a sky-wave link increases with decrease in the frequency of the transmission and, if the transmitted power is maintained at a constant level, the received field strength is inversely proportional to frequency. The *lowest useful frequency* (l.u.f.) is the lowest frequency at which a link with a given signal-to-noise ratio at the receive aerial can be established. The l.u.f. varies with time of day and year in a similar manner to the m.u.f.

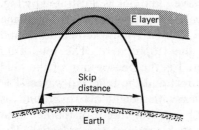

Fig. 6.10 Skip distance

Skip Distance

There is a minimum distance over which communication at a given frequency can be established by means of the sky wave. Usually, the frequency considered is the m.u.f. of the link. If an attempt is made to reduce this minimum distance by using a smaller angle of incidence, the wave will not be returned to earth by the E-layer but will pass through it. This minimum distance is known as the *skip distance* and is shown in Fig. 6.10. For a given frequency each of the ionospheric layers has its particular skip distance. It should be evident from the previous discussion that the higher the frequency of the wave the greater is the skip distance.

Multiple-hop Transmissions

When communication is desired between two points which are more than about 4000 km apart, it is necessary to employ two or more hops, as shown in Fig. 6.11. The sky wave is *refracted* in the ionosphere and returned to earth, and the downward wave is *reflected* at the surface of the earth to be returned skywards. The overall m.u.f. of a multi-hop link is the lowest of the m.u.f.s of the individual links.

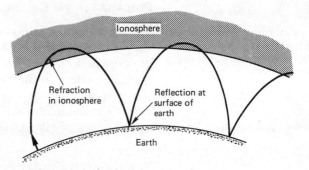

Fig. 6.11 Multi-hop transmission of sky wave

The number of hops that are possible depends upon both the transmitter power and the losses incurred at each ground reflection and ionospheric refraction. The main disadvantage of a multi-hop route is the likelihood of pronounced *selective fading*.

The Space Wave

At frequencies in the v.h.f., u.h.f. and s.h.f. bands, the range of the surface wave is severely limited and the ionosphere is unable to refract radio waves. Because the signal wavelength is short, the transmitting and receiving aerials can both be installed at a height of several wavelengths above earth. Then the

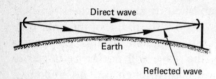

Fig. 6.12 The space wave

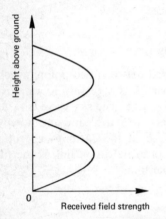

Fig. 6.13 Variation of field strength with height above ground at the receive end of a space wave radio link

space wave can be used for communication since its direct and reflected waves will not (always) cancel.

The principle of a space-wave radio link is illustrated by Fig. 6.12. The direct wave travels in a very nearly straight line path, slight refraction being caused by the temperature and water vapour gradients in the troposphere. The total field strength at the receiving aerial is the phasor sum of the field strengths produced by the direct and the reflected waves. The received field strength varies with height above ground as shown in Fig. 6.13. Obviously, careful choice of the height at which the receive aerial is to be mounted is essential. The maximum possible distance between the two aerials is somewhat greater than *line-of-sight* but, in practice, link lengths are shorter than this in order to improve the reliability of the system. Most links are some 25–40 km in length.

The direct wave must be well clear of any obstacles, such as trees and buildings which might block the path, and this factor will determine the necessary aerial heights.

The majority of point-to-point space wave radio systems are of considerably longer route length than 40 km and must of necessity require a number of *relay stations*. The use of relay stations will be discussed in Chapter 11.

Propagation via Communication Satellite

The basic principle of a communication satellite system is shown simply in Fig. 6.14. Since frequencies in the s.h.f. band are used in both directions of propagation, the ionosphere has negligible effect on the path of the radio waves, and so these travel in straight lines. This method of propagation can provide wideband multi-channel telephony systems over distances of thousands of kilometres with the utmost reliability.

Tropospheric Scatter Propagation

Another method of providing a number of radio-telephony channels over a long distance is known as *tropospheric scatter* and is illustrated by Fig. 6.15. A high-power radio wave is transmitted upwards from the earth and a very small fraction of the transmitted energy is *forward scattered* by the troposphere and directed downwards towards the earth. This occurs at frequencies above about 600 MHz, but particularly at 900 MHz, 2 GHz and 5 GHz. The forward-scattered energy is received by a high-gain aerial, often of the parabolic reflector type, to provide a reliable long-distance, wideband, u.h.f. radio link.The distance between the transmitting and receiving stations is usually in the range of 300 to 500 km and nearly always covers geographically hostile terrain, such as moun-

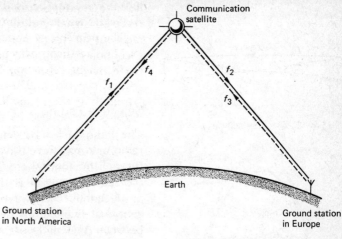

Fig. 6.14 An earth satellite communication system

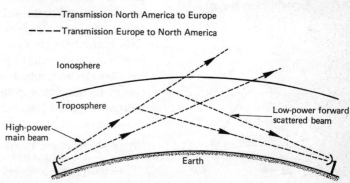

Fig. 6.15 Scatter propagation

tains, jungle or ocean. Since only a small fraction of the transmitted power arrives at the aerial, the system is very inefficient and demands the use of high-power transmitters and high-gain, low-noise radio receivers. For this reason a tropospheric scatter system is only provided when no other alternative is available.

Fading

Fading, or changes in the amplitude of a received signal, is of two main types: *general fading*, in which the whole signal fades to the same extent; and *selective fading*, in which some of the frequency components of a signal fade while at the same time others increase in amplitude.

General Fading

As it travels through the ionosphere, a radio wave is attenuated, but since the ionosphere is in a continual state of flux the attenuation is not constant, and the amplitude of the received signal varies. Under certain conditions a complete fade-out of signals may occur for up to two hours. With the exception of complete fade-outs, general fading can be combated by *automatic gain control* (a.g.c.) in the radio receiver.

Selective Fading

The radio waves arriving at the receiving end of a sky-wave radio link may have travelled over two or more different paths through the ionosphere (Fig. 6.16a). The total field strength at the receiving aerial is the *phasor* sum of the field strengths produced by each wave. Since the ionosphere is subject to continual fluctuations in its ionization density, the difference between the lengths of paths 1 and 2 will fluctuate and this will alter the total field strength at the receiver. Suppose, for example, that path 2 is initially one wavelength longer than path 1; the field strengths produced by the two waves are then in phase and the total field strength is equal to the algebraic sum of the individual field strengths. If now a fluctuation occurs in the ionosphere causing the difference between the lengths of paths 1 and 2 to be reduced to a half-wavelength, the individual field strengths become in antiphase and the total field strength is given by their algebraic difference.

The phase difference between the field strengths set up by the two waves is a function of frequency and hence the phasor sum of the two field strengths is different for each component frequency in the signal. This means that some frequencies may fade at the same instant as others are augmented; the effect is particularly serious in double-sideband amplitude-modulated systems because, if the carrier component fades to a level well below that of the two sidebands, the sidebands will beat together and considerable signal distortion will be produced.

Selective fading cannot be overcome by the use of a.g.c. in the receiver since this is operated by the carrier level only. Several methods of reducing selective fading do exist. For example, the use of frequencies as near to the m.u.f. as possible, the use of a transmitting aerial that radiates only one possible mode of propagation, the use of single-sideband or frequency-modulated systems, or the use of a specialized equipment as *Lincompex*. Selective fading of the sky wave is most likely when the route length necessitates the use of two or more hops. Suppose, for example, that a two-hop link has been engineered. Then, because of the directional characteristics of the transmitting aerial, there may well also be a three-hop path over which the transmitted energy is able to reach the receiving aerial.

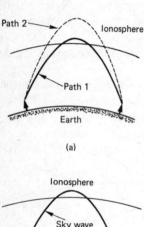

(a)

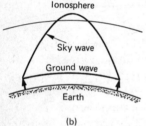

(b)

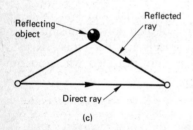

(c)

Fig. 6.16 Multi-path propagation

Selective fading can also arise with systems using the surface and space waves. In the daytime the D-layer of the ionosphere completely absorbs any energy radiated skywards by a medium-wave broadcast aerial. At night the D-layer disappears and any skywards radiation is returned to earth and will interfere with the ground wave, as shown in Fig. 6.16b. In the regions where the ground and sky waves are present at night, rapid fading, caused by fluctuations in the length of the sky path, occurs. This is why reception of medium-waveband broadcasts is much worse at night than in the daytime; it is minimized by the use of transmitting aerials having maximum gain along the surface of the earth and radiating minimum energy skywards.

Fig. 6.16c illustrates how multi-path reception of a v.h.f. signal can occur. Energy arrives at the receiver by a direct path and by reflection from a large object such as a hill or gasholder. If the reflecting object is not stationary the phase difference between the two signals will change rapidly and rapid fading will occur.

Use of the Frequency Bands

At frequencies in the v.l.f. and l.f. bands, aerials are very inefficient and high-power transmitters must be used. The radiated energy is vertically polarized and will propagate reliably (no fading) for thousands of kilometres using the surface wave or by means of multiple reflections between ionosphere and earth. Services provided in this band are ship-to-shore telegraphy, navigation systems, and sound broadcasting (l.f. band). In the m.f. band the range of the surface wave is limited to some hundreds of kilometres and the main use of the band is for sound broadcasting (647–1546 kHz). Also provided are ship telephonic and telegraphic links in, respectively, the bands 405–525 kHz and 1.6–3.8 MHz.

At high frequencies the main mode of propagation is the sky wave, the surface wave giving, if required, service for distances of up to the skip distance. The h.f. band is used for international point-to-point radio-telephony links on a number of sub-bands, for sound broadcasting, and for marine and aero mobile systems.

In the v.h.f. and higher bands the surface wave has a very limited range and the ionosphere (normally) does not return waves to earth. The modes of propagation used are therefore the space wave and, at certain frequencies in the s.h.f. band, the communication satellite. Scatter is also sometimes used. Services provided are sound broadcasting in the v.h.f. band (88.1–96.8 MHz), land, marine and aero mobile systems in the v.h.f. and u.h.f. bands, television broadcasting in the u.h.f.

band, and point-to-point multi-channel telephony systems in the u.h.f. and s.h.f. bands. (Details of the frequencies used are given in TSII.)

Exercises

6.1. (*a*) Which ionized regions are present in the atmosphere during a summer day? Give the approximate height of each of these regions. (*b*) What is meant by the following terms in connection with ionospheric propagation: (i) maximum usable frequency, (ii) gyro-frequency. (*c*) How is the maximum usable frequency related to (i) critical frequency, (ii) angle of incidence, (iii) density of ionization of the reflecting layer? (*C & G*)

6.2. (*a*) What is the meaning of the following terms: (i) maximum usable frequency, (ii) optimum traffic frequency? (*b*) How are (i) and (ii) related? (*c*) Explain how a radio wave incident on an ionized region is returned to earth by refraction. (*d*) How does the refraction vary with (i) frequency, (ii) electron density in the ionized region, (iii) angle of incidence? (*C & G*)

6.3. (*a*) Explain the mode of propagation whereby low radio frequencies can be used for world-wide communication. (*b*) How does field strength vary with the distance from the transmitter and with radiated frequency? (*c*) What are the advantages and disadvantages of low-frequency propagation? (*C & G*)

6.4. Explain the meanings of the following terms: (i) critical frequency, (ii) selective fading, (iii) surface wave, (iv) ground wave. (*b*) With what frequency band is each of the above normally associated? (*C & G*)

6.5. (*a*) Briefly describe the propagation of radio signals by means of the ground wave. (*b*) Why is the range of a broadcast transmitter using the ground wave limited even when the ground is lossless? (*c*) What is the name of the propagation mechanism whereby ground waves move around relatively small objects? (*d*) Does loss in the ground wave increase or decrease with increase in (i) conductivity, (ii) frequency, (iii) permittivity? (*C & G*)

6.6. (*a*) Which ionized regions are present in the atmosphere during a winter night? Give the approximate height of each of these regions. (*b*) What do you understand by the following terms in connection with ionospheric propagation: (i) skip distance, (ii) gyro-frequency? (*c*) What is the effect on the skip distance of an increase in (i) transmitted frequency, (ii) transmitted power, (iii) density of ionization of the reflecting layer? (*C & G*)

6.7. (*a*) Briefly describe what you understand by the terms (i) critical frequency, (ii) maximum usable frequency, (iii) optimum traffic frequency. (*b*) What is the relationship between (i) and (ii) in (*a*)? (*c*) Fig. 6.17 is a simplified diagram of an ionospheric region in the atmosphere: (i) explain how the radio wave incident on the ionized region represented in Fig. 6.17 is returned to earth when it is below the m.u.f.; (ii) why does a wave at a frequency above the m.u.f. penetrate the region? (iii) where is the maximum electron density in the ionospheric region shown in Fig. 6.17?

6.8. Explain how selective fading can arise on a long-distance short-wave radio link. How can its occurrence be minimized by a suitable choice of frequency? Briefly outline some radio transmission and reception techniques which reduce the effects of selective fading in (*a*) telephony, (*b*) telegraphy.

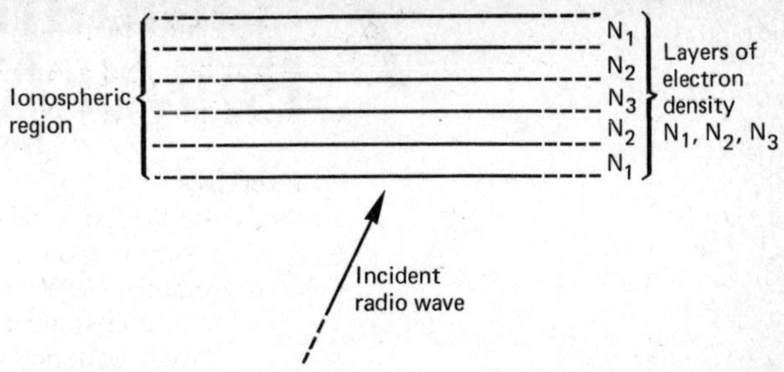

Fig. 6.17

6.9. Explain why v.h.f. and u.h.f. radio signals can be received beyond the line of sight distance from the transmitter. Explain why the height above ground of the aerial at the receiving end of the link is important.

6.10. Explain briefly, with the aid of sketches, how multipath interference occurs in the following types of radio transmission: (*a*) medium-frequency broadcasting, (*b*) high-frequency long-distance telephony, (*c*) v.h.f. television broadcasting. In each case state methods which are adopted to reduce the effects of each form of interference.

6.11. Explain how a long-distance wideband radio link can be established using tropospheric scatter in the u.h.f. band. A parabolic reflector aerial used in a 2 GHz tropospheric scatter system has a diameter of 6 m and radiates a power of 1 kW. Calculate the effective radiated power.

Short Exercises

6.12. List the ionospheric regions which are present in the upper atmosphere during a summer night. Give their approximate heights.

6.13. List the ionospheric regions which are present in the upper atmosphere during a winter day. Give their approximate heights.

6.14. Explain why as high a frequency as possible is used for a h.f. sky-wave transmission.

6.15. What is meant by the term gyro-frequency?

6.16. Explain the meanings of the terms ground wave and surface wave used in radio wave propagation.

6.17. For which frequency bands is the surface wave the main mode of propagation? Give typical ranges for each band.

6.18. Why does the surface wave suffer less attenuation over sea than over land?

6.19. Explain the meanings of the terms critical frequency, maximum usable frequency, and lowest useful frequency when used in radio propagation work.

6.20. What is meant by the term skip distance? Does the skip distance increase or decrease when the frequency is raised?

6.21. Give two reasons why a frequency as near the m.u.f. as possible is used for a sky-wave radio link.

6.22. Explain, with the aid of a diagram, why multi-hop sky-wave paths are prone to selective fading.

6.23. Does skip distance increase or decrease if (i) the frequency is raised, (ii) the transmitted power is increased, (iii) the electron density of the refracting layer is increased?

7 Radio-frequency Power Amplifiers

Introduction

Radio-frequency power amplifiers are used in radio transmitters to amplify the carrier frequency to the wanted power output level. The amplifiers are often expected to provide frequency multiplication at the same time as power amplification. For many radio transmitters the output power is of the order of tens, or perhaps hundreds, of kilowatts and such power levels are, at present, beyond the capacity of transistors. High-power transmitters therefore employ thermionic valves as the active device in the output stage. The earlier stages will probably use transistors. On the other hand, a low-power transmitter will be completely transistorized. Hence r.f. power amplifiers may use a transistor or a valve to provide amplification.

The choice between the triode valve and the tetrode valve must be made with due regard to a number of factors. The tetrode has a larger gain than the triode which means that a smaller input voltage is needed to develop a given output power. The anode-to-grid capacitance is considerably reduced by the screen grid, and usually the tetrode can be operated without *neutralization* circuitry. The triode will need to be neutralized, unless it is operated in the earthed grid configuration, to avoid positive feedback via its anode-grid capacitance. It is usually necessary to drive the triode valve into its grid current region in order to obtain a sufficiently large output power, and this practice results in distortion of the output waveform. The tetrode has the disadvantage that power is dissipated at its screen grid and so its overall efficiency is reduced; also some means of removing this heat may be necessary.

Radio-frequency power amplifiers are operated under either Class B or Class C conditions. Class B operation has a maximum theoretical efficiency of 78.5% but can be used to amplify an amplitude-modulated signal without the introduction of excessive distortion. The Class C amplifier has a higher efficiency but it cannot be used to amplify an amplitude-modulated signal. The extra efficiency provided by the Class C circuit is extremely important in a high-power application. For example, to obtain an output power of 100 kW requires an input power of 125 kW if the efficiency is 80%, but the input power must be 166 kW of the efficiency is reduced to 60%. The difference between the input and output powers is a waste of power and is dissipated in the valve in the form of heat and arrangements must be made to remove it. Even in the case of a low-power transmitter the utmost efficiency may still be important since such a transmitter may be battery operated.

The Class C Radio-frequency Power Amplifier

The basic circuit of a Class C radio-frequency power amplifier is given in Figs. 7.1a and b. A triode valve has been shown but equally a tetrode could be used. The difference between the two circuits shown lies in the way in which the anode-tuned circuit has been connected. In the SERIES-FEED circuit of Fig. 7.1a the tuned circuit is connected in series with the h.t. supply, while in the PARALLEL-FEED circuit of Fig. 7.1b the tuned circuit is connected in parallel with the valve and is isolated from the h.t. supply by the d.c. blocking capacitor C_2. The parallel-feed circuit uses another extra component, namely L_4, which stops r.f. currents entering the power supply instead of the tuned circuit C_3-L_5. In both circuits the valve is biased to operate under Class C conditions by the negative bias voltage V_b. Usually V_b is more than twice the cut-off voltage of the valve. The bias voltage is applied via inductor

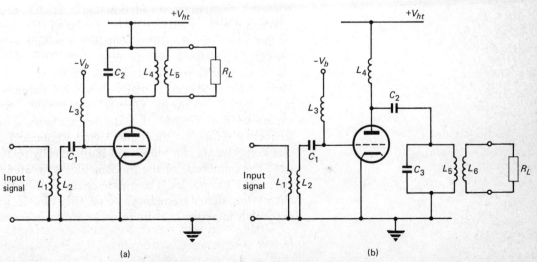

Fig. 7.1 Class C tuned power amplifiers: (a) series fed, (b) parallel fed

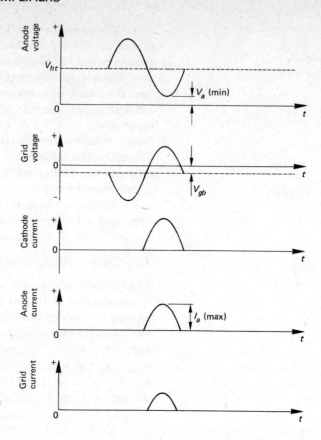

Fig. 7.2 Current and voltage waveforms in a Class C amplifier

L_3 to stop signal-frequency currents being shunted, via the bias supply, to earth. Capacitor C_1 prevents the bias voltage being shorted to earth by L_2. The parallel-feed arrangement ensures that the tuned circuit components are at zero d.c. potential which makes insulation and safety less of a problem.

When a sinusoidal voltage is applied to the input terminals of the amplifier, the valve will only conduct at the positive peaks of the signal voltage (Fig. 7.2). This means that the anode current flows as a series of pulses, each of which lasts for a time period smaller than one-half the periodic time of the input signal waveform. Clearly, the anode current is not sinusoidal but consists of a d.c. component, a fundamental frequency (equal to the input signal frequency) component, plus components at a number of harmonically related frequencies. The amplitude of the fundamental is greater than that of any of the harmonics. The anode circuit is tuned to be resonant at the signal frequency. A parallel-tuned circuit has its maximum impedance at its resonant frequency and this impedance, known as the DYNAMIC RESISTANCE R_d, is a pure resistance. At all other frequencies the impedance of a parallel-tuned circuit is much smaller and is not purely resisitive. The voltage developed across the anode circuit is therefore produced only by the fundamental (signal) frequency component of the anode current and is of sinusoidal waveform.

Very often with triodes (but not with tetrodes) the grid potential is taken positive with respect to the cathode at its peak positive half-cycles. This practice does result in the flow of grid current but it also produces anode current pulses of larger peak value than would otherwise be possible. Since the amplitude of the fundamental-frequency component of the anode current is proportional to the peak anode current, a larger output voltage is thus obtained.

The anode voltage of the valve is equal to the h.t. supply voltage minus the voltage developed across the anode tuned circuit and is in antiphase with the grid voltage. The current and voltage waveforms at various points in a Class C tuned amplifier are shown in Fig. 7.2. The following points should be noted:

(a) The anode current flows **whenever** the positive half-cycles of the input signal voltage make the grid potential less negative than the cut-off voltage of the valve.

(b) Grid current flows whenever the grid potential is positive.

(c) The minimum value of the anode voltage occurs at the same times as the positive peaks of the input signal voltage. It is necessary to ensure that at no time does the grid potential become more positive than the minimum anode voltage. If this should happen a large grid current will flow and damage the valve.

(d) The circuit has a high anode efficiency because anode current flows only at those times when the instantaneous anode voltage is at or near its minimum value. This reduces the power dissipated at the anode of the valve.

If the voltage of h.t. supply is increased, the peak value of the anode current pulses will increase in the same ratio and this, in turn, will increase the signal output voltage. Thus, in a Class C tuned amplifier the output voltage is *directly proportional* to the h.t. supply voltage.

Angle of Flow

The anode current of a Class C tuned power amplifier flows in a series of less-than-half sinewave pulses. The conduction time is expressed in terms of a parameter known as the ANGLE OF FLOW. The anode current flows whenever the total grid voltage is less negative than the cut-off voltage of the valve. Thus, referring to Fig. 7.3, which shows, in more detail than Fig. 7.2c, the grid voltages existing in the circuit, θ is the angle of anode current flow. In this figure V_{co} is the cut-off voltage of the valve and V_b is the bias voltage applied. Also shown is the angle ϕ of grid current flow, always $\phi < \theta$.

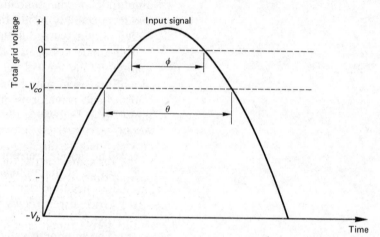

Fig. 7.3 Angle of flow

The angle of flow is always less than 180° ($\theta = 180°$ gives Class B conditions), the actual value chosen being a compromise between the conflicting requirements of anode efficiency and power output. A reduction in the angle of flow reduces the power output of the amplifier but increases its efficiency. Generally, the angle of anode current flow is chosen to be somewhere in the region of 120°.

When θ is approximately 120° the anode current flows for only one third of each cycle and the current waveform can be considered, without the introduction of undue error, to be of triangular shape. This is a convenient assumption to be able to make since it allows the mean value of the anode current to be easily calculated using equation (7.1):

$$I_{a(mean)} = \frac{I_{a(peak)}}{2} \times \frac{\theta°}{360°} \qquad (7.1)$$

The mean value of the anode current is the direct current which is taken from the h.t. power supply.

EXAMPLE 7.1

The anode current pulses in a Class C tuned amplifier are of approximately triangular waveform with a peak value of 3 A and an angle of flow of 100°.
Calculate the mean value of the anode current.

Solution
From equation (7.1)

$$I_{a(mean)} = \frac{3}{2} \times \frac{100}{360} = 0.417 \text{ A} \qquad (Ans.)$$

Power Relationships

The d.c. power supplied by the h.t. power supply to the amplifier is equal to the product of the h.t. supply voltage V_{ht} and the mean value of the anode current:

$$P_{dc} = V_{ht}I_{a(mean)} \tag{7.2}$$

Some of this power is dissipated at the anode of the valve and is known as the *anode dissipation*. The remainder of the input power is converted into a.c. power and is delivered to the anode tuned circuit. A small amount of this power is dissipated in the resistance of the tuned circuit inductance but the rest is passed onto the load to provide the power output of the amplifier. Normally, the power lost in the tuned circuit is small and will be neglected in this book. The a.c. power P_{ac} delivered to the anode tuned circuit, i.e. the a.c. output power, is equal to the square of the r.m.s. value of the fundamental frequency component $I_{a(f)}$ of the anode current times the dynamic resistance R_d of the tuned circuit:

$$P_{ac} = I_{a(f)}^2 R_d \tag{7.3}$$

The ANODE EFFICIENCY η of a Class C amplifier is the ratio P_{ac}/P_{dc} expressed as a percentage

$$\eta = \frac{P_{ac}}{P_{dc}} \times 100\% \tag{7.4}$$

Alternatively, the power output of the circuit can be expressed in terms of the anode, and h.t. supply voltages. The fundamental frequency component of the anode current develops a voltage $V_L \sin \omega t$ across the tuned circuit, where $V_L = I_{a(f)}R_d$.
The output power can therefore be written as $(V_L/\sqrt{2})^2/R_d$.
The anode voltage V_a of the valve is $V_a = V_{ht} - V_L \sin \omega t$.
When $\sin \omega t = 1$ anode voltage is at minimum value $V_{a(min)}$.

Then $V_L = V_{ht} - V_{a(min)}$

Therefore,

$$P_{ac} = \frac{(V_{ht} - V_{a(min)})^2}{2R_d} \tag{7.5}$$

EXAMPLE 7.2.

In a Class C tuned power amplifier the anode current is of approximately triangular waveform of peak value 4.6 A and angle of flow 120°. If the anode efficiency of the circuit is 75% and the h.t. supply voltage is 1 kV calculate the output power.

Solution

$$P_{dc} = \frac{4.6}{2} \times \frac{120°}{360°} \times 1000 = 766.7 \text{ W}$$

Therefore,

$$P_{ac} = \eta P_{dc} = 0.75 \times 766.7 = 575 \text{ W} \qquad (Ans.)$$

Earthed-grid Operation of a Triode Class C Amplifier

When a triode valve is used in the common-cathode connection, feedback of radio-frequency energy from the anode (output) circuit to the grid (input) circuit via the anode-grid capacitance of the valve will take place. This *positive feedback* will produce instability and possibly unwanted oscillations. To overcome the instability problem it will be necessary to use extra circuitry, known as NEUTRALIZATION, to cancel or neutralize the unwanted feedback. In modern transmitters the

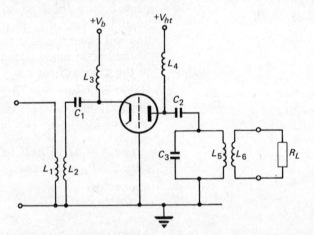

Fig. 7.4 Earthed grid Class C tuned power amplifier

need for neutralization is overcome by connecting the triode in the earthed-grid configuration (Fig. 7.4). The earthed grid now acts as a screen between the cathode and anode electrodes which reduces the cathode-anode capacitance to a very small value. Now unwanted feedback from the output (anode) to input (cathode) circuit is at very low level and generally neutralization circuitry is not needed.

Grid Bias

The valve must be biased to operate under Class C conditions and the required bias voltage can be obtained from a separate

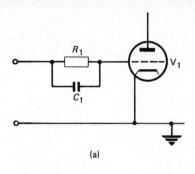

(a)

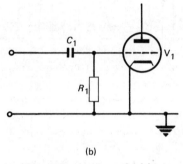

(b)

Fig. 7.5 Leaky-grid bias

power supply as shown in the circuits of Fig. 7.1*a* and *b*. An alternative arrangement which can be used instead of, or as well as, a power supply is the LEAKY-GRID bias circuit. The action of a leaky-grid bias circuit, two versions of which are shown in Fig. 7.5, has been described elsewhere [EII] but, very briefly, the required bias voltage is developed across capacitor C_1 by the flow of grid current. The use of leaky-grid bias has the disadvantage that if, for some reason, the input signal voltage is removed, the bias voltage will disappear. The valve may then pass a very large anode current and quite possibly suffer damage. To prevent such an occurrence a small cathode resistor can be fitted or some fixed-bias can also be provided.

Effect of Loading

The anode tuned circuit must be tuned to the required frequency of operation and then possess sufficient selectivity to be able to discriminate against the harmonic content of the anode current. These requirements are satisfied by using an inductor of high Q-factor and a low-loss capacitor. In addition, the tuned circuit must provide a suitable load impedance for the valve and must also transfer the power from the anode circuit to the load. The effective resistance of the tuned circuit is equal to its dynamic resistance in parallel with a coupled resistance. The magnitude of this coupled resistance depends upon both the coupling between the anode circuit and the load and the load impedance itself. By suitable adjustment of this coupling the optimum load for the valve can be obtained.

The efficiency with which energy is transferred from the anode circuit to the load depends upon the ratio (loaded Q)/(unloaded Q). This means that the unloaded Q-factor should be high but should fall to a low value once the load is connected. The Q-factor cannot be permitted to fall to too low a figure however or insufficient discrimination against harmonics will be provided. Typically, the loaded Q-factor is about 12.

Class B Tuned Power Amplifiers

The Class C tuned power amplifier has the advantage of a very high anode efficiency but it can only be used when the input signal voltage is of constant amplitude. If an amplitude-modulated signal were applied to the amplifier, considerable distortion would be introduced. This reason for this is illustrated by Fig. 7.6 which shows a 75% modulated wave applied to a Class C biased valve characteristic. The peaks of the modulation envelope produce anode current pulses of varying peak value, but during the troughs of the envelope the grid

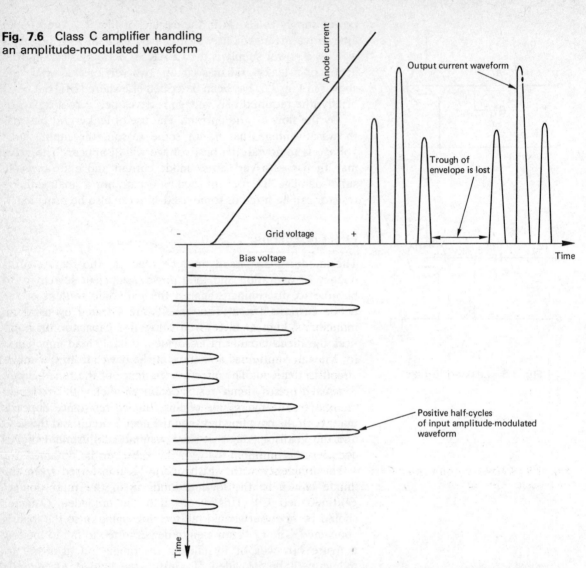

Fig. 7.6 Class C amplifier handling an amplitude-modulated waveform

voltage is unable to drive the valve into conduction. This form of distortion can only be prevented by ensuring that all positive half-cycles of the input signal voltage, no matter how small, cause anode current to flow. This effectively means that the bias voltage should be reduced until it is equal to the valve's cut-off voltage, i.e. the valve must be operated under Class B conditions.

In practice, the mutual characteristics of a valve tend to be non-linear for the lower values of anode current and it is usual, in order to minimize distortion caused by this non-linearity, to use *projected grid bias*, as shown in Fig. 7.7. The valve is not now operated under true Class B conditions since a small anode current will flow for zero input signal voltage, but little error is introduced in calculation by assuming that

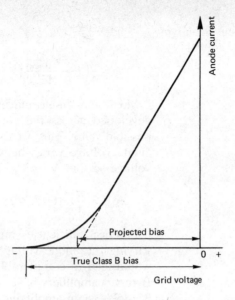

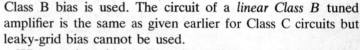

Fig. 7.7 Projected Class B bias

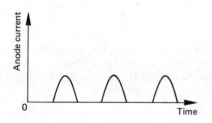

Fig. 7.8 Anode current waveform in a Class B amplifier

Class B bias is used. The circuit of a *linear Class B* tuned amplifier is the same as given earlier for Class C circuits but leaky-grid bias cannot be used.

When a sinusoidal voltage is applied to a Class B tuned amplifier, the anode current will flow as a series of half-sinewave pulses (Fig. 7.8) of peak value $I_{a(max)}$. The fundamental frequency component of this current will develop a voltage V_L across the anode tuned circuit. The mean value of this current is $I_{a(max)}/\pi$ and the peak value of its fundamental frequency component is $I_{a(max)}/2$. A number of components at other frequencies are also present in the current waveform but they will have little contribution to make to the output voltage.

The d.c. power taken from the h.t. supply is

$$P_{dc} = \frac{I_{a(max)}}{\pi} \times V_{ht}$$

The a.c. power output is

$$\frac{I_{a(max)}}{2\sqrt{2}} \times \frac{V_L}{\sqrt{2}}$$

Therefore

$$P_{ac} = \frac{I_{a(max)}}{2\sqrt{2}} \times \frac{(V_{ht} - V_{a(min)})}{\sqrt{2}}$$

$$= \frac{I_{a(max)}}{4}(V_{ht} - V_{a(min)}) \qquad (7.6)$$

The anode efficiency η of a Class B tuned amplifier is the ratio $(P_{ac}/P_{dc}) \times 100\%$. Therefore

$$\eta = \frac{I_{a(max)}}{4}(V_{ht} - V_{a(min)}) \times \frac{\pi}{I_{a(max)}\, V_{ht}}$$

$$= \frac{\pi}{4}\left(1 - \frac{V_{a(min)}}{V_{ht}}\right) \times 100\% \tag{7.7}$$

Maximum anode efficiency is obtained when the voltage developed across the anode tuned circuit has its maximum possible value. This occurs when the valve is driven so that its anode voltage varies between zero and twice the h.t. supply voltage so that $V_L = V_{ht}$. Then $V_{a(min)} = 0$ and

$$\eta_{max} = \frac{\pi}{4} \times 100\% = 78.5\% \tag{7.8}$$

Practical efficiencies must fall short of this figure because varying the anode voltage over such a wide range of values would lead to considerable signal distortion. Generally, Class B tuned amplifiers have an anode efficiency in the region of 35–45% when amplifying a sinusoidal signal. The anode efficiency will rise when an amplitude-modulated wave is amplified by about 10%.

EXAMPLE 7.3

A Class B tuned power amplifier operates with a h.t. voltage of 1000 V and a peak anode current of 6 A. If the effective dynamic resistance of the anode tuned circuit is 200 Ω calculate (a) the output power and (b) the anode efficiency of the amplifier.

Solution
(a) The fundamental frequency component of the anode current has a peak value of $6/2 = 3$ A.
Therefore,

$$P_{ac} = \left(\frac{3}{\sqrt{2}}\right)^2 \times 200 = 900 \text{ W} \qquad (Ans.)$$

(b) The mean anode current $= 6/\pi$ A and the power taken from the h.t. supply is $6000/\pi$ W. Therefore,

$$\eta = \frac{900}{6000/\pi} \times 100\% = 47.1\% \qquad (Ans.)$$

The Class B tuned power amplifier is often operated in the earthed grid configuration both to avoid neutralization and because of its improved linearity.

EXAMPLE 7.4

In the amplifier of Example 7.3 the input signal is amplitude modulated to a depth of 60%. Calculate the anode efficiency of the amplifier.

Solution
From equation (1.9),

Total output power $= 900\ (1+\tfrac{1}{2}0.6^2) = 1062\ \text{W}$

The d.c. power taken from the supply is unchanged and so

$$\eta = \frac{1062}{6000/\pi} \times 100\% = 55.6\% \qquad (Ans.)$$

Transistor Tuned Power Amplifiers

The transistor versions of the Class B and Class C tuned power amplifiers are used in low-power radio transmitters, in particular for those in mobile systems. The circuit of a transistor Class C tuned power amplifier is shown in Fig. 7.9. Leaky-base bias [EII] is provided by R_1 and C_2 with L_3 acting to prevent

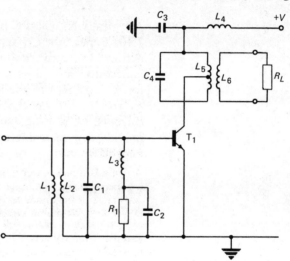

Fig. 7.9 Transistor Class C tuned amplifier

signal-frequency currents passing to earth via C_2. Alternatively, a separate bias supply could be used. The collector tuned circuit C_4-L_5 is series fed and coupled by mutual inductance to the load R_L. The anode circuit is tapped to obtain the optimum load impedance for T_1. Inductor L_4 prevents r.f. currents passing into the power supply and C_3 is a decoupling component. The circuit of a Class B transistor tuned power amplifier is very similar to the circuit of Fig. 7.9 but a separate base bias supply voltage must be provided to give slight forward bias; otherwise the transistor base/emitter junction of the transistor will develop a self-bias that would give Class C operation.

When a transistor tuned power amplifier is designed for operation in the v.h.f. or u.h.f. bands, it is usually expected to work between $50\,\Omega$ source and load impedances. This impedance is not a suitable value for a v.h.f. or u.h.f. transistor to work in between, and often input and output T and/or π matching networks are used to obtain more convenient values.

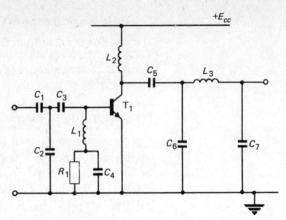

Fig. 7.10 Transistor Class C tuned amplifier with π coupling to load

A typical v.h.f. Class C power amplifier circuit is shown in Fig. 7.10. Capacitors C_1, C_2 and C_3 form the input T matching network and convert, at the design frequency, the 50 Ω source impedance into the source impedance required by the transistor. C_4 and R_1 provide leaky-base Class C bias and inductors L_1 and L_2 are r.f. chokes which prevent the passage of r.f. currents. C_5 is a d.c. blocking component and C_6, C_7 and L_3 provide a π matching network which converts the 50 Ω load impedance into the correct impedance for T_1 to work into.

EXAMPLE 7.5

A Class C transistor tuned power amplifier operates from a 30 V collector supply. If the collector dissipation of the transistor is 1.2 W and the mean collector current is 0.1 A determine (a) the a.c. output power, (b) the collector efficiency of the amplifier.

Solution
(a) $P_{dc} = 30 \times 0.1 = 3 \text{ W}$
    Therefore,

$$P_{ac} = P_{dc} - \text{collector dissipation} = 3 - 1.2 = 1.8 \text{ W} \qquad (Ans.)$$

(b) $\eta = \dfrac{1.8 \times 100\%}{3} = 60\%$ (Ans.)

Frequency Multipliers

The anode or collector current of a Class C tuned amplifier flows as a series of less-than-half sinewave pulses and contains components at the input signal frequency and at harmonics of that frequency. If the anode or collector tuned circuit is tuned to be resonant at a particular harmonic of the input signal frequency, the voltage developed across the load will be at that harmonic frequency. The angle of flow should be chosen as $180°/n$ where n is the order of the harmonic required. Thus, if a frequency tripler is to be designed, the angle of flow should be 60°. The higher the order of harmonic selected by the anode (collector) tuned circuit the smaller will be the angle of flow and thus the smaller will be the output power. In practice, the frequency multiplication obtained is rarely in excess of 5. When a larger multiplying factor is wanted two or more frequency multipliers are connected in cascade.

Anode-modulated Class C Tuned Amplifiers

The output voltage of a Class C tuned power amplifier is directly proportional to the h.t. supply voltage. If the h.t. supply voltage is increased by, say, 50% the amplitude of the voltage developed across the anode tuned circuit also increases by 50%. This means that if the h.t. supply voltage is caused to vary in sympathy with a modulating signal voltage, the a.c. voltage developed across anode tuned circuit will be *amplitude modulated.*

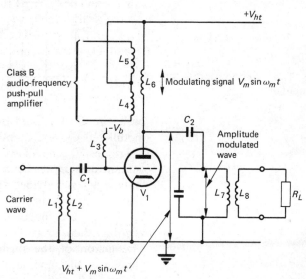

Fig. 7.11 Anode-modulation of a Class C amplifier

The h.t. voltage applied to the anode of a Class C tuned power amplifier can best be varied by introducing the modulating signal into the anode circuit by means of a transformer as shown in Fig. 7.11. The total voltage applied to the anode of V_1 is the sum of the h.t. supply voltage V_{ht} and the modulating signal voltage $V_m \sin \omega_m t$ which appears across inductor L_6. The maximum voltage applied to the anode of V_1 is $V_{ht} + V_m$ and the minimum voltage is $V_{ht} - V_m$. The depth of modulation of the output voltage waveform depends upon the relative voltages of the h.t. supply and the modulating signal. If, for example, $V_m = V_{ht}/2$ the depth of modulation will be 50%.

The modulating signal voltage is produced by an audio-frequency power amplifier. Sometimes this circuit may be operated under Class A conditions but, for high-power applications, it is always a Class B push-pull amplifier [EIII]. The

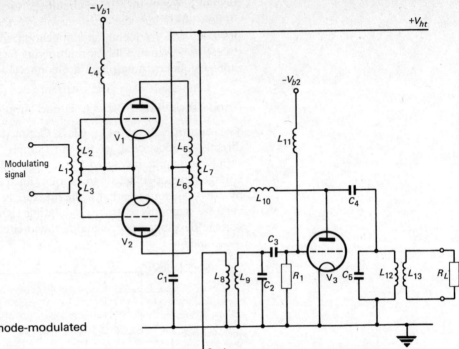

Fig. 7.12 The anode-modulated Class C amplifier

basic circuit of a Class B modulator and its associated anode-modulated Class C amplifier is shown in Fig. 7.12.

Triodes V_1 and V_2 are connected in push-pull and operated very nearly under Class B conditions; a small forward bias voltage V_{b1} is provided via L_4. L_4, L_{10} and L_{11} are r.f. chokes. The modulating signal is first applied to inductor L_1 and, since L_2 and L_3 are the two halves of a centre-tapped secondary winding, then applied in antiphase to the grids of V_1 and V_2. The action of the push-pull circuit produces an amplified version of the modulating signal voltage across L_7. Thus, the effective h.t. voltage applied to the Class C amplifier is V_{ht} + $V_m \sin \omega_m t$. Leaky-grid bias, augmented by the bias supply V_{b2}, is provided by C_3 and R_1 while C_4 functions merely as a d.c. blocking component.

When there is no modulating signal applied to the circuit, the voltage applied to the anode of V_3 is the h.t. supply voltage V_{ht} and the output voltage developed across the load is of constant amplitude. The power output of the amplifier is the unmodulated carrier power. This means that the carrier power is equal to the d.c. power supplied to the Class C amplifier times its anode efficiency.

When a modulating signal is applied, the h.t. voltage supplied to V_3 is varied in accordance with its characteristics and an amplitude-modulated waveform is produced across the load. The power developed across the load resistance has increased by an amount equal to the power contained in the upper and lower sidebands. This extra power must have been supplied by the modulator stage. Hence in an anode modulated Class C amplifier:

(a) The carrier power output is equal to the d.c. power supplied to the Class C amplifier times the efficiency of the amplifier.

(b) The sideband power output is equal to the power provided by the Class B modulator times the efficiency of the Class C stage. The output power of the Class B modulator is equal to the d.c. power supplied to the modulator times the efficiency of the modulator. This means that the sideband power is equal to the d.c. power input to the Class B stage times the product of the anode efficiencies of the Class B and Class C stages.

EXAMPLE 7.6

A Class C anode-modulated amplifier uses a Class C stage with an anode efficiency of 75% and a Class B stage whose anode efficiency is 50%. The sinusoidally modulated output wave form has a depth of modulation of 50% and a total power of 1520 watts. Calculate the d.c. power supplied to (a) the Class C stage, (b) the Class B stage.

Solution
From equation (1.9)

Total power $P_t = 1520 = P_c \ (1 + \tfrac{1}{2}0.5^2)$

$P_c = 1520/1.125 = 1351$ W

Total sidefrequency power $= 1520 - 1351 = 169$ W

(a) D.C. power supplied to the Class C stage is

$P_{dc(C)} = 1351/0.75 = 1801.3$ W   (*Ans.*)

(b) Sidefrequency power supplied to the Class C stage is

$P_{SF} = 169/0.75 = 225.3$ W

Therefore, d.c. power supplied to the Class B stage is

$P_{dc(B)} = 225.3/0.5 = 450.6$ W   (*Ans.*)

The depth of modulation of the output amplitude-modulated waveform depends upon the amplitude of the modulating signal voltage introduced in series with the h.t. supply voltage of the Class C stage. For 100% modulation the modulating signal voltage must be equal to the h.t. voltage so that the effective h.t. voltage for the Class C stage varies between 0 and $2V_{ht}$.

The grid bias voltage must be provided wholly, or at least mainly, by means of a leaky-grid bias circuit. If, using fixed bias, the circuit is adjusted to function correctly as a Class C amplifier when the effective h.t. voltage is twice the d.c. supply voltage, then during the troughs of the modulation cycle the

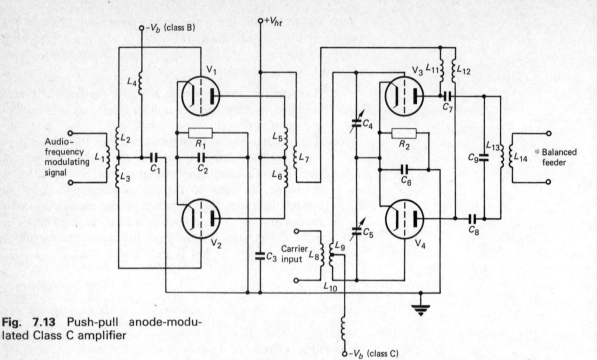

Fig. 7.13 Push-pull anode-modulated Class C amplifier

instantaneous anode voltage will be so small that an excessive grid current flows. If leaky-grid bias is used, the bias voltage will vary throughout the modulation cycle and maintain the correct bias for the instantaneous h.t. voltage.

Sometimes to increase the power output a push-pull Class C stage is used; a typical circuit is shown in Fig. 7.13.

Collector-modulated Class C Tuned Amplifiers

A d.s.b. amplitude-modulated wave can be generated by collector-modulating a transistor Class C tuned amplifier. The modulating signal is introduced in series with the collector supply voltage to modulate the voltage applied to the amplifier. The basic circuit of a collector-modulated Class C amplifier is given by Fig. 7.14. Transistors T_1 and T_2 form a push-pull amplifier which operates very nearly in Class B, a small forward bias being given by the potential divider $R_1 + R_2$. The modulating signal voltage is developed across L_6 and so appears in series with the collector supply voltage E_{cc}. Transistor T_3 operates in Class C with leaky-base bias provided by R_3-C_3 and uses a π-type network to couple the output voltage to the load.

If a high depth of modulation is required, one stage of modulation is likely to prove insufficient. Then the penultimate Class C amplifier must also be modulated and Fig. 7.15 shows a possible circuit. The collector supply voltage to both the

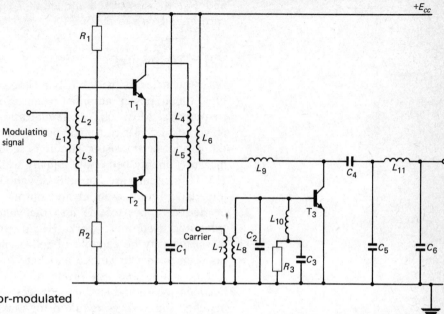

Fig. 7.14 The collector-modulated Class C amplifier

penultimate T_3 and the final T_4 Class C amplifiers passes through the inductor L_3. Both Class C amplifiers are therefore collector modulated by the push-pull modulator T_1/T_2. The Class C stages are shown with T and π input and output coupling networks. Capacitors C_3, C_6 and C_8 are d.c. blocking components, C_1 and C_5 are decoupling capacitors, and inductors L_6 and L_9 act as the collector load impedances. Finally, the necessary base bias voltages are provided by resistors R_2

Fig. 7.15 Collector modulation of final and penultimate Class C transistor stages

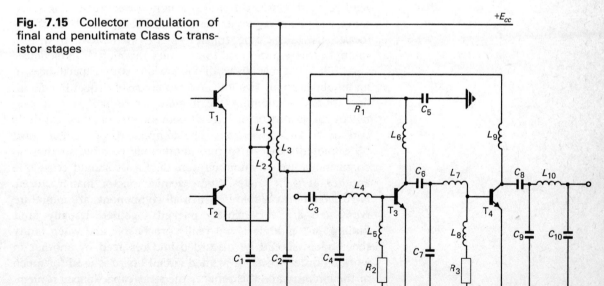

and R_3, the associated inductors L_5 and L_8 preventing r.f. currents passing to earth.

V.H.F. Techniques

At frequencies in the v.h.f. band the design and construction of a tuned power amplifier is more difficult than at lower frequencies. Most v.h.f. communication transmitters are low-powered and use transistor power amplifiers; in this section the use of a transistor circuit is assumed but most of the measures mentioned apply equally well to a valve circuit.

V.H.F. circuits are normally designed to work between $50\,\Omega$ impedances and so input and output coupling networks are needed to convert $50\,\Omega$ into the values between which the transistor itself must work. Because of internal feedback a transistor may be capable of self-oscillation when connected between particular source and load impedances. A transistor will be stable and not prone to oscillate if it is connected between source and load impedances somewhat lower than the matched impedances required for maximum power gain. Because of the impedance values concerned, the input and output coupling circuits are always either T or π networks as shown in Figs. 7.14 and 7.15. Of course, the transistor used should be a v.h.f. type designed to have minimum internal feedback, adequate current gain, and minimum noise at the operating frequency. The inductance of the lead joining the emitter electrode to the emitter pin can have an appreciable reactance at the higher end of the v.h.f. band and particularly in the u.h.f. band. This reactance will cause negative feedback to be applied to the circuit and also a non-power dissipating resistance to appear across its base-emitter terminals; both effects reduce its gain. Some transistors are manufactured with a multiple-wire emitter lead to minimize this emitter inductance.

At v.h.f. the reactances of the various stray capacitances in an amplifier circuit are low and can adversely affect the circuit operation. To minimize the magnitudes of the stray capacitances the layout of the circuit components must be carefully considered and carried out. The components of a circuit must be mounted as close to one another as possible so that all connections are of minimum length. Leads should cross one another at right angles. Some semiconductor manufacturers have *modules* available in which all components are miniature types and are very closely packed together. Usually fault finding on a module is not really practicable and when faulty the module should be discarded and replaced by another in working order. Often, a printed circuit board is used for much of the circuitry and this reduces the stray capacitance problem. When a lead passes from one side to the other of the chassis, or a metal screening can, a feedthrough capacitor should be used.

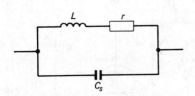

Fig. 7.16 Equivalent circuit of an inductor

The components used in a v.h.f. amplifier must all be chosen with some care. All resistors possess both self-inductance and self-capacitance, all inductors possess resistance and capacitance, and all capacitors have both inductive and resistive components. If a component is used at a frequency higher than it was designed for, it may not provide the electrical characteristics expected. For example, Fig. 7.16 shows the electrical equivalent circuit of an inductor, and C_S represents its self-capacitance. At some particular frequency the inductor will be self-resonant and act like a high-value resistor; at higher frequencies the component will have an effective *capacitance*. Clearly, an inductor must be operated at frequencies well below its self-resonant frequency.

The various stages in a circuit should be positioned in as near a straight line as possible and be individually screened. To avoid unwanted couplings between different parts of a circuit arising from currents flowing in the chassis and/or the earth line, all earth connections *in one stage* should be made to a common point. A separate common point should be used for each stage.

Exercises

7.1. (*a*) A point-to-point high power communication receiver has its r.f. output stage modulated by a push-pull class B modulator. Sketch the circuit of the modulator and r.f. output stage. (*b*) The power output of an a.m. anode-modulated transmitter is 1245 W. The efficiency of the final stage is 60%. If the modulation depth is 0.7 and the efficiency of the modulator is 50%, calculate (i) the modulation power supplied to the anode of the final stage, (ii) the anode dissipation of the modulator stage. Ignore losses in the modulation transformer. (*C & G*)

7.2. (*a*) Draw the circuit diagram of a push-pull Class B modulator with its associated Class C r.f. amplifier. (*b*) A push-pull Class B modulator is used to modulate sinusoidally a push-pull Class C r.f. amplifier. The maximum anode dissipation of the r.f. amplifier is 250 W and its anode efficiency is 75%. The Class B modulator has an anode efficiency of 60% and a maximum anode dissipation of 200 W. (i) Calculate the maximum modulated r.f. output from the Class C amplifier, (ii) What is the maximum modulation power that the modulator can supply to the r.f. amplifier? (iii) What is the maximum depth of modulation? (*C & G*)

7.3. Explain briefly how a Class C amplifier may be used as a frequency multiplier. Illustrate your answer by waveforms of collector voltage, collector current, base voltage, and base current. Indicate how the harmonic number of a multiplier influences the choice of (*a*) angle of collector current flow and (*b*) collector load impedance.

7.4. Draw the circuit diagram of the output stage of a high-frequency telephony transmitter using high-power modulation, and explain the operation of the circuit. Briefly discuss the relative merits of high-power and low-power modulation.

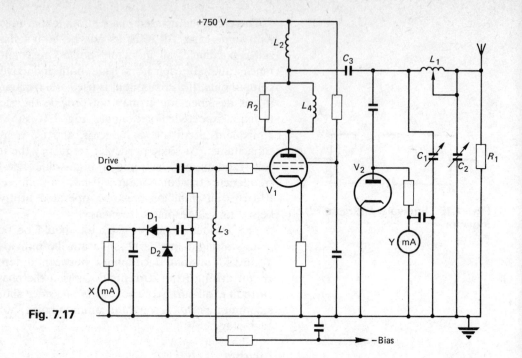

Fig. 7.17

7.5. Draw the circuit diagram of a Class C tuned power amplifier which uses a p-n-p transistor as the active device and is designed to operate between source and load impedances of 50 Ω. Explain the operation of the circuit.

7.6. Fig. 7.17 is the circuit diagram of the output stage of an s.s.b. transmitter.
(*a*) Why is fixed bias used instead of automatic bias? (*b*) The π filter minimizes the radiation of harmonics. What is the cause of these harmonics? (*c*) What is the purpose of the meter at Y? (*d*) How is the aerial ammeter connected? (*e*) How is the π filter adjusted to match the amplifier to the aerial? (*f*) What is the purpose of the meter at X? (*g*) What would be the effect of an open-circuited D_2? (*h*) Why is resistor R_2 connected across the inductor L_4? (*i*) What is the purpose of R_1? (*j*) What factors determine the peak voltage at the aerial output terminal?

(*C & G*)

7.7. (*a*) Make a list of the components of the collector-modulated Class C amplifiers of Fig. 7.14 and for each one state its function. (*b*) Describe the operation of the circuit.

7.8. Fig. 7.18 shows the block diagram of an anode-modulated Class C tuned amplifier.
(*a*) The efficiencies of the Class B and Class C stages are, respectively, 48% and 66%, and the output waveform has a depth of modulation of 60%. If the total output power is 11 kW calculate the d.c. power inputs to the two stages.
(*b*) The output from such a circuit has a total power of 30 kW and a modulation depth of 70%. If the Class C stage has an efficiency of 70% calculate the efficiency of the Class B modulator.

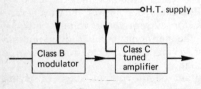

Fig. 7.18

Short Exercises

7.9. A Class C r.f. amplifier has an angle of flow of 130°. Sketch the variation with time of its collector and base voltage over one cycle of the sinusoidal applied voltage. Show on your sketch the base bias voltage and the base cut-off voltage.

7.10. With the aid of suitable mutual characteristics explain the difference between Class A, Class B and Class C operation in an amplifier stage.

7.11. The collector current in a Class C amplifier has a peak value of 5 A, an angle of flow of 120° and is of triangular waveform. Calculate the mean value of the current.

7.12. Draw the circuit of a series-fed collector-modulated Class C tuned amplifier.

7.13. Why are triodes in Class C tuned amplifiers often driven into the grid current region while tetrodes are not?

7.14. The d.c. power taken from the 2000 V h.t. supply of a tuned amplifier is 1000 W. Determine the peak value of the anode current pulses if the amplifier is operated under (i) Class B and (ii) Class C conditions.

7.15. The a.c. power output of a tuned amplifier is 500 W. If the anode efficiency of the circuit is 70% calculate its anode dissipation.

7.16. Draw the circuit diagram of an earthed-grid Class B tuned power amplifier. State the purpose of each component shown.

7.17. Explain briefly, with the aid of a sketch, why amplitude-modulated signals cannot be handled by a Class C tuned amplifier.

7.18. List the functions of a tuned power amplifier.

7.19. (*a*) Why does the collector current of a Class C amplifier flow as a series of less-than-half sinewave pulses? (*b*) Why is it that the output voltage waveform is not severely distorted?

8 Radio Transmitters

Introduction

The purpose of any radiocommunication system is to transmit intelligence from one point to another; the communication may be unidirectional as in the case of sound and television broadcasting or it may be bi-directional as with most radio-telephony systems. At the transmitting end of the system the signal must modulate a suitable carrier frequency to translate the signal to the allocated part of the frequency spectrum, and then be amplified to the necessary transmitted power level.

In the v.h.f. and u.h.f. bands both amplitude and frequency-modulation transmitters are used but in the lower frequency bands only amplitude modulation finds application. Sound broadcast transmitters use d.s.b. amplitude modulation but radio-telephony systems use either single or independent-sideband operation. High-frequency communication transmitters must be able to alter frequency rapidly as ionospheric propagation conditions change in order to maintain a reliable service. Modern h.f. transmitters are designed to be self-tuning to facilitate the frequency-changing process.

Amplitude-modulation Transmitters

In an amplitude-modulated radio transmitter, the carrier wave is generated by a high-stability crystal oscillator or a frequency synthesizer, and then amplified, and perhaps frequency multiplied, before it is applied to the aerial feeder. At some stage in the process the carrier is amplitude-modulated by the information signal. The modulation can be carried out when the carrier is at a low level or after it has been amplified to a high power level and transmitters are broadly divided into two classes, namely low-level and high-level, for that reason.

The block diagram of a HIGH-LEVEL TRANSMITTER is given by Fig. 8.1. The carrier frequency is generated by the

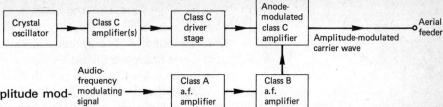

Fig. 8.1 High-level amplitude modulation transmitter

crystal oscillator and is amplified to the level necessary to fully drive the output stage by a number of Class C tuned r.f. amplifiers. One or more of these amplifiers may be operated as frequency multipliers. The modulating signal is amplified by the Class A a.f. amplifier and then applied to the Class B modulator. The output of the modulator is connected in the anode circuit of the final Class C stage and amplitude-modulates the amplified carrier wave. The frequency stability of the transmitter is dependent upon the stability of the crystal oscillator. Since this is generally improved if overtone operation of the crystal can be avoided, it is often the practice to operate the crystal at its fundamental frequency and use the appropriate frequency multiplication to obtain the required carrier frequency. A modern practice which can sometimes be used is the use of a crystal oscillator at a higher frequency than the required carrier and to use frequency division; an improved frequency stability can then be achieved.

The advantage of the high-level method of operating a radio transmitter is that high-efficiency Class C tuned amplifiers can be used throughout the r.f. section. The disadvantage is that the a.f. modulating signal must be amplified to a high power level if it is to adequately modulate the carrier. This demands the use of a high-power Class B a.f. amplifier and this, mainly because of the output transformer requirements, is an expensive item of equipment. When the method is used in a low-power transistorized mobile transmitter, this disadvantage tends to disappear. High-level modulation is used for d.s.b. amplitude-modulated sound broadcast, and v.h.f/u.h.f. mobile transmitters.

The LOW-LEVEL method of operating an amplitude-modulation transmitter is shown in Fig. 8.2. The carrier voltage receives little, if any, amplification before it is modulated by the signal. The amplitude-modulated wave is then amplified

Fig. 8.2 Low-level amplitude modulation transmitter

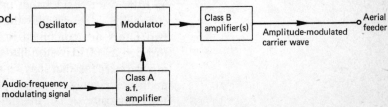

by one or more linear Class B r.f. power amplifiers to the wanted output power level. In general, Class C tuned amplifiers cannot amplify amplitude-modulated waveforms without generating excessive distortion. Some modern low-level transmitters use Class C amplifiers with envelope negative feedback to reduce the distortion. The low-level transmitter does not require a large a.f. modulating power, which simplifies the design of the a.f. amplifiers. On the other hand compared with high-level operation its overall efficiency is much lower because Class B amplifiers are used in place of Class C circuits.

The majority of d.s.b. amplitude-modulation transmitters use the high-level method of modulation mainly because of the greater efficiency offered. The low-level method of operation is widely used for s.s.b. and i.s.b. transmitters, the modulation process being carried out in a separate *drive unit* (Fig. 8.3). The audio-frequency signal is applied to the drive unit and is there converted into an s.s.b. or i.s.b. waveform which is then passed on to the main transmitter. In the main transmitter the s.s.b./i.s.b. signal is amplified to the required power level and translated to the appropriate part of the frequency spectrum. The h.f. communication i.s.b. transmitters used in the U.K. for international radio-telephony links use a standard drive unit, and all variations in transmitted frequency and/or power are provided by the main transmitter.

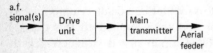

Fig. 8.3 Low-level s.s.b./i.s.b. transmitter

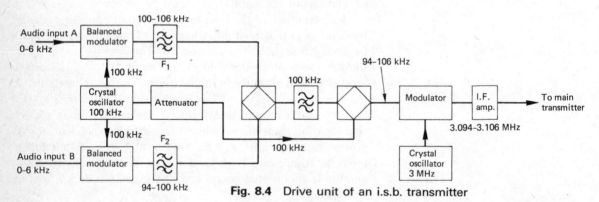

Fig. 8.4 Drive unit of an i.s.b. transmitter

The block diagram of the standard DRIVE UNIT is given in Fig. 8.4. The audio input signal to a channel is applied to a balanced modulator together with a 100 kHz carrier signal, and the wanted sideband, upper for channel A and lower for channel B, is selected by the channel filter. The outputs of the two filters are combined in the hybrid coil and then passed through a 100 kHz stop filter. This filter is provided to remove any carrier leak that may be present at the outputs. If required at the receiver a low-level *pilot carrier* is then reinserted into the composite signal, the function of this pilot carrier being to operate the automatic gain control and automatic frequency control circuitry in the receiver. The 94–106 kHz i.s.b. signal is then translated to the band 3.094–3.106 MHz by modulation of the 3 MHz carrier. The output of the drive unit is passed to the main radio transmitter for radiation at the required frequency somewhere in the band 4–30 MHz.

Since the required bandwidth for a speech circuit is approximately 3 kHz, each 6 kHz sideband is capable of accommodating two speech channels. Telegraphy signals require an even narrower bandwidth, and so several telegraph channels can be accommodated in the place of one or more speech channels. The method generally employed for obtaining four 3 kHz telephony channels for application to the channel A and B input terminals of the i.s.b. drive unit is shown in Fig. 8.5. Fig. 8.5*a* shows the transmitting-end equipment while Fig. 8.5*b* shows the equipment required at the receiving end. The equipment is not usually located at the site of the radio transmitter but is installed at the radio telephony terminal, and each pair of channels, in the band 250–6000 Hz, is sent to the transmitter over a four-wire line circuit.

Fig. 8.5 Method of deriving four a.f. channels for transmission over an i.s.b. system

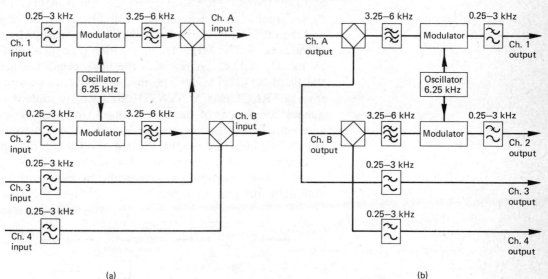

(a)   (b)

The block diagram of the drive unit for a s.s.b.s.c. transmitter is given, with typical frequencies at each point, in Fig. 8.6; transmitters of this kind are used in maritime mobile systems. A pilot carrier version of the unit is also possible, the pilot being added to the s.s.b. signal in the same way as in Fig. 8.4.

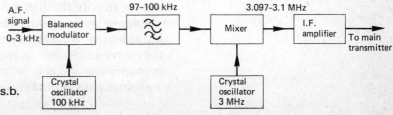

Fig. 8.6 Drive unit of an s.s.b. transmitter

Communication transmitters operating in the h.f. band must be capable of rapid and frequent changes in operating frequency as ionospheric propagation conditions vary. If the tuning and loading process is carried out manually (described later in this chapter) it may take 20 minutes or so to complete the procedure and for this reason many modern transmitters are *self-tuning*. With a self-tuning transmitter the operator has merely to reset some dials at a control position and the

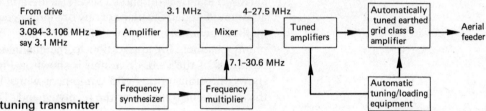

Fig. 8.7 A self-tuning transmitter

tuning/loading process is carried out automatically in about 20–30 seconds. The simplified block schematic diagram of a SELF-TUNING TRANSMITTER is shown in Fig. 8.7. The i.s.b. output of the drive unit shown in Fig. 8.4 is first amplified and then frequency translated to its allocated carrier frequency in the band 4–27.5 MHz. The translation process is carried out by mixing the i.s.b. signal with the appropriate frequency in the band 7.1–30.6 MHz. The mixing frequencies are derived from a FREQUENCY SYNTHESIZER. The difference frequency component of the mixer output waveform is selected, and amplified, by several cascaded tuned amplifiers, to the voltage required to drive the earthed-grid Class B output stage to give the rated output power. The tuned amplifiers and the Class B stage are tuned automatically by motor-driven variable inductors and capacitors.

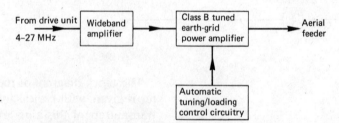

Fig. 8.8 Another self-tuning transmitter

Another version of the self-tuning transmitter (Fig. 8.8) employs a wideband amplifier which does not require tuning in order to amplify the signal provided by the drive unit to the level necessary to operate the tuned power output amplifier. The wideband amplifier operates over the entire frequency band covered by the transmitter. The drive unit for this transmitter must be able to produce the i.s.b. or s.s.b. signal at the desired frequency of operation and hence differs from the unit of Fig. 8.4 in that its output frequency is not constant. The output frequency of the drive unit, and thus of the transmitter, is determined by a frequency synthesizer. When the transmitted frequency is to be altered, the synthesizer is set to generate the appropriate frequency which, after mixing with the i.s.b./s.s.b. signal, produces a signal at the new wanted frequency. The tuning and loading of the tuned Class B output stage is automatically adjusted to the correct value in about 30 seconds.

The power output of a self-tuning h.f. transmitter is several kilowatts, typically 20 kW.

A self-tuning transmitter may be called upon to work at any frequency in the band 4–27.5 MHz. Since this is a wider bandwidth than a rhombic or a log-periodic aerial can efficiently work over, some kind of *aerial switching* is often used

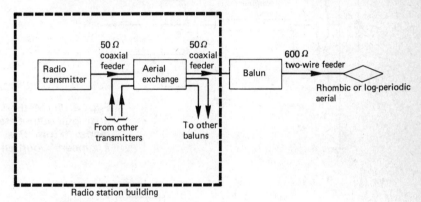

Fig. 8.9 Use of an aerial exchange

and Fig. 8.9 illustrates a possible arrangement. A number of transmitters are connected to the *aerial exchange*; this is a switching array that makes it possible for any of the transmitters to be switched to any particular aerial.

Output Stages

The output stage of a h.f. transmitter must be designed to satisfy a number of requirements which are listed below:

(*a*) It must transfer the wanted output power to the aerial feeder with the utmost efficiency.

(*b*) It must have sufficient selectivity to discriminate against the unwanted harmonic components of the anode current, but not against the side frequencies of the signal.

(*c*) It should operate in a stable and linear manner.

(*d*) Tuning the stage to the required operating frequency and optimizing its coupling to the aerial feeder should be as easy and rapid a process as possible. Often this process is carried out automatically.

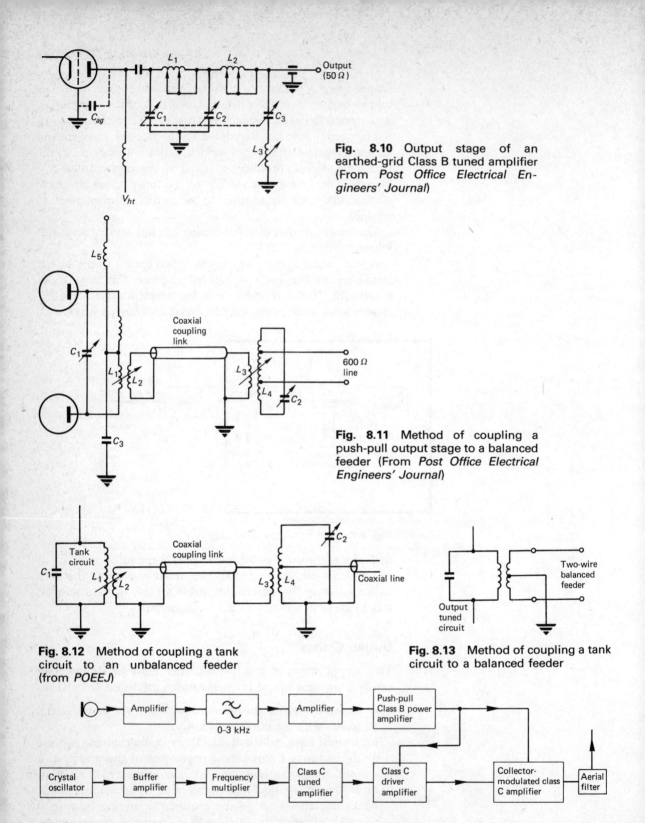

Fig. 8.10 Output stage of an earthed-grid Class B tuned amplifier (From *Post Office Electrical Engineers' Journal*)

Fig. 8.11 Method of coupling a push-pull output stage to a balanced feeder (From *Post Office Electrical Engineers' Journal*)

Fig. 8.12 Method of coupling a tank circuit to an unbalanced feeder (from *POEEJ*)

Fig. 8.13 Method of coupling a tank circuit to a balanced feeder

Fig. 8.14 Amplitude-modulation v.h.f. transmitter

Sound broadcast transmitters usually employ a Class C biased output stage but communication transmitters, handling i.s.b./s.s.b. signals, use an output stage that is operated under Class B conditions and Fig. 8.10 shows a typical circuit. The circuit is tuned to the wanted frequency in the band 4–27.5 MHz by the π-type network consisting of C_1, C_2 and L_1. Unwanted second harmonics of the selected frequency are suppressed by the series-tuned circuit C_3-L_3 which provides a low-resistance path to earth at its resonant frequency. Optimum coupling to the 50 Ω coaxial feeder is obtained by suitable adjustment of the value of inductor L_2. All the variable components are motor-driven and automatically adjusted when the operating frequency of the transmitter is to be altered.

Two other methods of coupling an output stage to a feeder are shown in Figs. 8.11 and 8.12. Fig. 8.11 shows how a Class B push-pull output stage would be coupled to a 600 Ω twin feeder. The 600 Ω impedance of the feeder is changed to the load impedance value required by the valves by the settings of the tapping points on inductor L_4. The coupling between the output stage and the feeder is optimized by adjustment of the mutual inductance coupling between L_1 and L_2 and between L_3 and L_4. The coupling arrangement is tuned to the required frequency by means of capacitors C_1 and C_2. L_5 and C_3 are power supply decoupling components.

Fig. 8.12 shows how an anode tuned circuit could be connected to an unbalanced coaxial feeder. The components of the coupling network have similar functions to those shown in Fig. 8.11. Coupling an output tuned circuit to a balanced twin feeder can be achieved in a simpler manner as shown by Fig. 8.13. The secondary winding of the output transformer is centre-tapped to ensure that both conductors are at the same potential relative to earth.

V.H.F. Mobile Transmitters

The output power of most v.h.f. transmitters is only a few tens of watts and so completely solid state equipments can be designed. The block schematic diagram of a v.h.f. amplitude-modulated transmitter is shown in Fig. 8.14. The voltage generated by the microphone is amplified and then band-limited by the 3 kHz cut-off low-pass filter. The band-limited signal is amplified to the power level necessary for it to collector-modulate the driver and output stages of the transmitter. Although not shown in the figure, a transmitter of this type often has the facility for switching different crystal oscillators into circuit to permit operation at different frequencies. The carrier frequency can be generated directly at frequencies up to about 150 MHz or so using an overtone mode of the crystal. However, overtone crystal operation tends to be more expensive and of poorer frequency stability than the use of a lower-frequency crystal oscillator followed by one or more frequency multipliers. Mobile v.h.f. channels are positioned very close to one another in the frequency spectrum and it is important that a transmitter radiates little, if any, power at other frequencies. To ensure the adequate suppression of spurious frequencies an aerial filter is connected between the output stage of the transmitter and the aerial.

The frequency at which a radio transmitter operates must be maintained constant to within internationally agreed limits to avoid interference with adjacent (in frequency) channels. In the case of s.s.b. and i.s.b. systems, the suppressed carrier must be re-inserted at the receiver with the correct frequency. This requirement will clearly be made harder if the carrier frequency at the transmitter is not constant. The oscillator from which the transmitter carrier frequency is derived must be of stable frequency, both short- and long-term. If the operating frequency of a transmitter is frequently changed, a variable-frequency oscillator of some kind must be fitted but it will then be difficult to achieve the desired frequency stability.

The highest frequency stability is obtained with a *crystal oscillator*. At frequencies near the higher end of the h.f. band it is customary to employ a crystal oscillator operating at a low frequency and then to use one or more stages of frequency multiplication to obtain the wanted transmitted frequency. A crystal oscillator is a fixed-frequency circuit, and if a transmitter is to operate at different frequencies it will be necessary to switch different crystals into circuit. Many modern transmitters use a technique known as *frequency synthesis* to derive all the necessary frequencies. Typically, the frequency stability of a high-frequency transmitter is ± 1 part in 10^6, i.e. ± 1 Hz if the carrier frequency is 10 MHz.

Frequency Synthesis

A frequency synthesizer is an equipment which derives a large number of discrete frequencies, singly or simultaneously, from an accurate high-stability crystal oscillator source. Each of the derived frequencies has the accuracy and stability of the source. A synthesizer may cover a wide frequency band, for example any frequency at 100 Hz increments between 4 MHz and 10 MHz. The use of a frequency synthesizer in h.f. communication transmitters has already been mentioned; many radio receivers also use frequency synthesis and this is mentioned in the following chapter. In most equipments the wanted frequency is selected by means of a number of decade switches but in the latest circuits digital control is used.

There are two main methods available for the operation of a synthesizer generally known as the *direct* and *indirect* methods. With the DIRECT METHOD a required output frequency is obtained by a process of frequency multiplication, mixing and filtering. The stability and accuracy of the output frequency is set by the reference (crystal oscillator) source but the method has two disadvantages which have led to its unpopularity for use in modern systems. These disadvantages are that (*a*) the mixing process generates spurious frequencies, (*b*) a number of relatively costly filters are needed.

Most modern equipments use the INDIRECT METHOD of frequency synthesis. With this method the required output frequency is derived from a voltage-controlled oscillator whose accuracy is maintained by phase-locking the oscillator to a

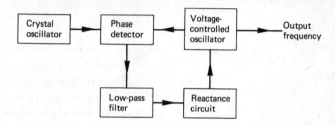

Fig. 8.15 Indirect frequency synthesis

standard frequency. The principle of an indirect frequency synthesizer is illustrated by Fig. 8.15. The phase detector is a circuit which, when voltages at the same frequency are applied to its two input terminals, produces a direct output voltage whose magnitude and polarity is proportional to the *phase difference* between the two input voltages. The direct output voltage is applied to a voltage-variable reactance (e.g. a varactor diode) which is connected as a part of the frequency-determining circuit of the oscillator. The effective capacitance of the reactance circuit is varied by the direct control voltage in the direction necessary to reduce the phase error of the voltage-controlled oscillator.

Any tendency for the voltage-controlled oscillator to change frequency is opposed by the phase-control loop. As the frequency starts to drift, a control voltage is generated by the phase detector which changes the reactance in the direction necessary to correct the frequency drift. The control voltage must be free from alternating components, produced by noise or distortion, otherwise the output frequency will not be stable. This is the reason for the inclusion of the low-pass filter in the loop.

There are various methods by which the output of the voltage-controlled oscillator can be used to derive all the wanted frequencies. One possible arrangement is shown by

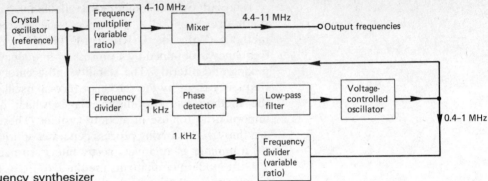

Fig. 8.16 Frequency synthesizer

Fig. 8.16. The output of a highly stable crystal oscillator is fed into a frequency-multiplication circuit which can produce an output frequency at any integral number of megahertz between 4 and 10. The selected frequency is mixed with the output of the voltage-controlled oscillator and the *sum* frequency is selected. The voltage-controlled frequency can be set to any frequency in the range 0.4–1 MHz. and is maintained accurately at this frequency by the phase-locked loop. The outputs of the crystal oscillator and the voltage-controlled oscillator are both divided down to 1000 Hz and applied to the phase detector.

Another version of a frequency synthesis equipment is shown in Fig. 8.17. Any frequency in the band 4–8 MHz is made available by generating five different frequencies f_1, f_2, etc. and combining them by repeated mixing and filtering. Each of the five frequencies is produced by the phase lock circuit of Fig. 8.17*b* which shows the frequencies used for the f_3 decade (10–100 kHz). The crystal oscillator can operate at any one of ten frequencies in the band 3.555 to 3.645 MHz by switching different crystals into circuit.

Frequency-modulation Transmitters

Frequency modulation is used for sound broadcasting in the v.h.f. band, for v.h.f. and u.h.f. mobile systems, and for wideband s.h.f. radio-relay systems. Radio-relay transmitters are covered in Chapter 11.

The block schematic diagram of an f.m. sound broadcast transmitter is shown in Fig. 8.18. The modulating signal is applied to the input terminals of a varactor-diode modulated *L-C* oscillator to frequency-modulate the carrier. The frequency-modulated output waveform is then amplitude-limited to remove any amplitude modulation introduced by the modulator, before it is multiplied and amplified to the specified output power and frequency. Typical figures are

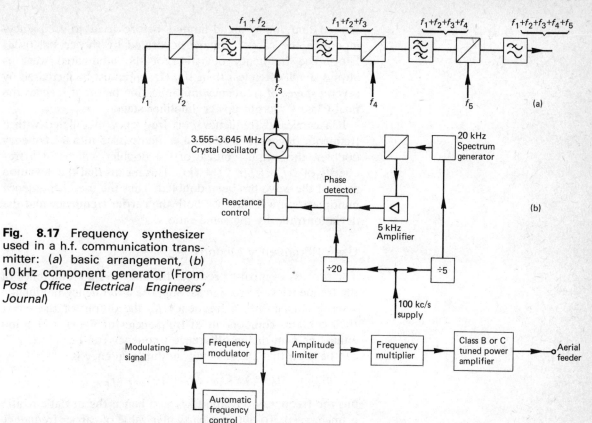

Fig. 8.17 Frequency synthesizer used in a h.f. communication transmitter: (a) basic arrangement, (b) 10 kHz component generator (From *Post Office Electrical Engineers' Journal*)

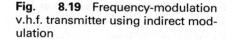

Fig. 8.18 Frequency-modulation v.h.f. transmitter

91.3 MHz and 10 kW. The centre (unmodulated) carrier frequency radiated by the transmitter must be very stable, typically ± 1 kHz per year, and since the inherent frequency stability of an *L-C* oscillator is inadequate automatic frequency control must be applied.

Many transmitters, particularly those used in narrowband mobile systems, use the indirect method of frequency modulating a carrier because of the improved frequency stability then provided. The block diagram of a typical narrowband f.m. communication transmitter is given by Fig. 8.19. The microphone output voltage is amplified, integrated, amplitude-

Fig. 8.19 Frequency-modulation v.h.f. transmitter using indirect modulation

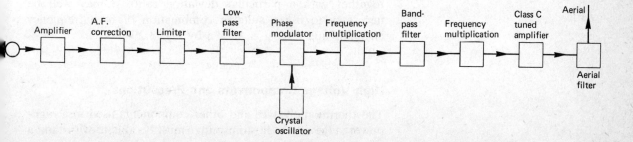

limited and finally band-limited before used to frequency-modulate the carrier voltage generated by the crystal oscillator. The frequency deviation of the modulated wave is always small, often less than 100 Hz, and must be increased by several stages of frequency multiplication before it reaches the final Class C output power amplifier stage.

If a carrier at frequency f_c is frequency modulated with a frequency deviation of kf_d and is then passed into a frequency doubler, the output voltage of the doubler will be at a frequency of $2(f_c \pm f_d)$ or $2f_c + 2kf_d$. This means that the deviation ratio of the wave has been doubled. Thus the use of frequency multiplication will increase both the carrier frequency and the deviation ratio by the same ratio.

Use of Frequency Mixing

The process of mixing produces the sum and the difference of the frequencies of two signals. Suppose a frequency-modulated wave is mixed with a frequency f_0; the output of the mixer then contains components at frequencies of $f_0 \pm (f_c \pm kf_d)$ and either the sum or the difference frequency can be selected. It can be seen that the selected output frequency is

$$\text{either} \quad (f_0 + f_c) \pm kf_d \quad \text{or} \quad (f_0 - f_c) \pm kf_d$$

but the frequency deviation kf_d and hence the deviation ratio is unchanged. To obtain a particular value of carrier frequency

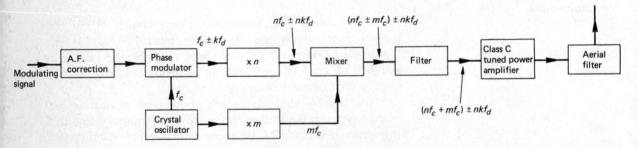

Fig. 8.20 Method of obtaining particular values of carrier frequency and deviation ratio

together with a particular deviation ratio, it may well be necessary to use a suitable combination of both frequency changing and mixing as shown by Fig. 8.20.

High Voltage Components and Precautions

The thermionic valves and other components used in a high-power stage of a radio transmitter must be able to withstand a voltage of several thousands of volts and so special high-voltage types must be used. The maximum permissible voltage which can be safely applied across a resistor is limited because of the danger of a dielectric breakdown. Any such breakdown appears in the form of sparking. Another factor which must be considered is the power rating of a resistor. If this is exceeded, excessive heat will be dissipated within the component and its resistance will change, possibly by a significant amount. Ceramic carbon resistors are probably the best type to use since they have voltage ratings of 20 kV or more. Similarly the voltage rating of a capacitor is determined by the need to ensure that the dielectric between the plates does not break down.

The design of a high-voltage inductor is constrained by the need to avoid the insulation (very likely air) between adjacent turns breaking down because of the electric field across it. This means that the turns must be spaced well apart from one another and probably be self-supporting. Also, the inductor in a series-feed circuit will have a current of several amperes flowing in it and must be manufactured using a conductor of large cross-sectional area. Often the power dissipated within a high-voltage inductor is so large that the component must be cooled by blowing cool air around it. In all cases high-voltage components must not be allowed to become too hot and the equipment design must ensure adequate ventilation.

The valve in the output stage, and perhaps the penultimate stage also, of a high-power transmitter may be called upon to dissipate several kilowatts of power at its anode, and in the case of a tetrode at its screen grid also. The anode will become extremely hot because of the power dissipated at it and this heat must be removed to keep the temperature of the anode within acceptable limits. When the anode dissipation is not very large, say less than 1.5 kW, sufficient heat can be removed by fabricating the copper anode as an integral part of the valve's envelope and relying on direct radiation. When the anode dissipation is larger than 1.5 kW the rate of removal of heat must be speeded up by passing cooling air, or water, around the anode. In modern transmitters, cooling of the anode is achieved using a vaporization technique. The basic idea of a water vapour cooling system is that cooling water is converted into steam by the heat of the anode, this steam is removed and condensed back into water, and then the water is re-circulated past the anode.

Because of the very high voltages present at various points in a high-power transmitter, various precautions must be taken to ensure the safety of the persons required to carry out tests or repairs on the equipment. All high-voltage points in the transmitter are mounted inside interlocking cabinets or cages. Entry within a cage can only be made by following a procedure which ensures that the panel giving access to the interior of the equipment can only be opened *after* the high-voltage has been removed. This is generally arranged by feeding in the power supplies via an isolating unit that is interlocked with an earthing switch. The keys necessary to unlock the panel can only be obtained after the isolating unit has been operated to disconnect the power supply and earth the equipment. Usually, different parts of the transmitter are housed inside different cabinets, for example one cabinet might contain the power supply circuitry and another the r.f. power output stage. When work on the equipment is completed it is necessary to follow the reverse procedure, i.e. all entry panels must be replaced and locked and the keys restored *before* the power supplies can be switched on again.

Exercises

8.1. (*a*) Describe, with the aid of a block diagram, the drive and r.f. stages of a high-power independent-sideband h.f. transmitter. How is crosstalk between sidebands minimized? (*b*) Compare and contrast independent-sideband operation with (i) double sideband, (ii) single sideband operation.   (*C & G*)

8.2. (*a*) Briefly discuss the considerations which enter into the design of the output stage of an h.f. transmitter. (*b*) Give two reasons why each aerial at an h.f. transmitting station is not permanently associated with a transmitter. (*c*) Explain briefly why a transmitter must be matched to an aerial feeder.
(*C & G*)

8.3. Fig. 8.21 is the block diagram of a marine f.m. transmitter. (*a*) State the purpose of the limiter in the microphone amplifier. (*b*) What is the purpose of the audio-frequency corrector? (*c*) How is the stability of the frequency generator maintained for different channels? (*d*) Explain briefly the function of the modulator. (*e*) If the frequency generator has an 8.7 MHz output what is the frequency band required for speech signals? (*f*) State briefly how an amplifier may be used as a frequency multiplier. (*g*) If the power amplifier uses a self-biased valve operated in Class C, what would be the effect on its anode current of the failure of a previous stage? (*h*) Why is a vertical aerial used for transmission?   (*C & G*)

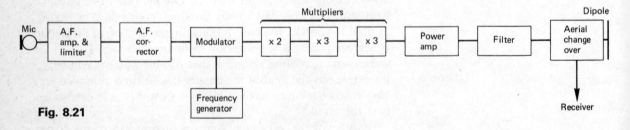

Fig. 8.21

8.4. Draw the block diagram of the drive and r.f. stages of a high-power short-wave independent-sideband transmitter. (*b*) Describe briefly the operation of this transmitter, describing how the independent sidebands are combined. (*c*) Draw the spectrum of the signal at three representative points in your diagram taking care to distinguish between the sidebands in each. (*C & G*)

8.5. (*a*) Draw the block diagram of a medium-frequency broadcast transmitter. (*b*) What is the class of operation of each stage? (*c*) How is frequency control obtained? (*d*) For such a transmitter, what is a typical (i) frequency range, (ii) frequency stability (iii) power delivered to the aerial? (*C & G*)

8.6. (*a*) What do you understand by the terms (i) high-level modulation, (ii) low-level modulation? Illustrate your answers with block diagrams of amplitude-modulation transmitters. (*b*) What class of operation is usual for the final stage amplifiers for each of these types of transmitter? (*c*) What do the following abbreviations stand for when used in conjunction with amplitude-modulation transmissions: (i) s.s.b., (ii) d.s.b., (iii) d.s.b.s.c., (iv) i.s.b.? (*d*) Using sketches of a sinusoidal modulating signal and a sinusoidal carrier signal illustrate the waveform of a modulated signal using (i) d.s.b., (ii) d.s.b.s.c., (iii) s.s.b. types of modulation. (*C & G*)

8.7. (*a*) Draw the block diagram of an amplitude-modulated high-power telephony transmitter for a point-to-point system covering the 4 MHz to 27.5 MHz band. (*b*) Describe briefly the operation of this transmitter, explaining how rapid frequency changing is facilitated. (*c*) Why is regular frequency changing necessary and about how long does it take in a modern transmitter? (*d*) Using a sketch, show how, in a modern system, a transmitter of this type is switched from one aerial to another. (*C & G*)

8.8. With the aid of a circuit diagram explain the operation of a v.h.f. telephony transmitter using high-power modulation. (*C & G*)

8.9. (*a*) Using a fully labelled block diagram, explain the principle of operation of the Armstrong method of obtaining wideband frequency modulation. (*b*) In a particular Armstrong modulator the side-frequencies produced by a 3 kHz sinusoidal modulating signal are each 6 dB down on the amplitude of the 30 kHz sub-carrier. If the final carrier is radiated at a nominal 90 MHz, calculate the frequency deviation of this transmission. (*C & G*)

8.10. Discuss the advantages and disadvantages of the following as the drive unit for a transmitter: (*a*) variable-frequency oscillator, (*b*) crystal-controlled oscillator, (*c*) frequency synthesizer. (*C & G*)

8.11. (*a*) What is meant by the terms (i) deviation ratio, (ii) modulation index as applied to frequency modulation transmissions? (*b*) With aid of a circuit diagram show how, by means of a varactor diode, a 5 MHz oscillator could be frequency modulated by a 2 kHz audio signal. (*c*) The output of the oscillator is converted to 90 MHz by (i) frequency multiplication, (ii) mixing with output of an 85 MHz oscillator. Assuming the deviation produced by the varactor diode to be 3 kHz, deduce the final deviation in each case. (*C & G*)

8.12. (*a*) What is meant by the term "frequency synthesis"? (*b*) Use a block diagram to illustrate how frequency synthesis is used in a modern radio telephony transmitter. (*c*) In a certain synthesizer a $2(n-2)$ divider chain output is equal to a 500 kHz reference frequency. If the input to the dividers is the difference between ten times the reference frequency and the synthesized frequency, what value of n should be selected to derive a 158 kHz output? (*C & G*)

8.13. (*a*) Using circuit diagrams to illustrate your answers show how (i) a telephony transmitter with a Class B push-pull output stage would be coupled to a 600 Ω transmission line, (ii) a transmitter output tank circuit would be connected to a coaxial line aerial feeder. Show how matching and tuning are facilitated. (*b*) Briefly explain how a high-power transmitting valve is protected in the event of a feeder or aerial failure. (*C & G*)

Short Exercises

8.14. Why is there a need of high-efficiency in the final stage of a radio transmitter?

8.15. What is the function of a tuned power amplifier in a radio transmitter?

8.16. (*a*) Why does an h.f. communication transmitter need to change frequency fairly often? (*b*) Why does an h.f. broadcast transmitter remain at the same frequency?

8.17. Why is frequency stability necessary in a radio transmitter? How is it obtained in (i) an h.f. communications transmitter, (ii) a v.h.f. f.m. communication transmitter?

8.18. List the requirements of the output stage of a high-frequency radio transmitter.

8.19. Refer to the block diagram of a v.h.f. a.m. transmitter given in Fig. 8.14 and answer the following questions. (*a*) Why is the modulating signal band-limited? (*b*) Why are two stages of r.f. power gain collector modulated? (*c*) How is the transmitted frequency changed? (*d*) Why is an aerial filter fitted?

8.20. What is meant by frequency synthesis and why is it used in modern h.f. transmitters?

8.21. Draw the block diagram of an i.s.b. drive unit and a low-level main transmitter and show how the transmitter can be switched to operate with one of several aerials.

8.22. (*a*) Why is a limiter often used in an f.m. transmitter? (*b*) Why is a phase modulator often used?

9 Radio Receivers

Introduction

The functions of a radio receiver are to select the wanted signal from all those signals picked up by the aerial, to extract the information which has been modulated on to the wanted signal, and then to amplify the signal to the level necessary to operate the loudspeaker or other receiving device. A radio receiver may be designed to receive sound broadcast signals using d.s.b. amplitude modulation or using frequency modulation; for use with land, maritime, or aero-mobile systems using amplitude or frequency modulation; or for use in a multi-channel point-to-point radio link. For reasons discussed elsewhere [RSII] radio receivers are of the *superheterodyne* type.

The Superheterodyne Radio Receiver

In a superheterodyne radio receiver the wanted signal frequency is converted into a constant frequency—known as the intermediate frequency—at which most of the gain and the selectivity of the receiver is provided.

The basic block diagram of a superheterodyne radio receiver is shown in Fig. 9.1. The wanted signal, at frequency f_s, is passed, together with many other unwanted frequencies, by

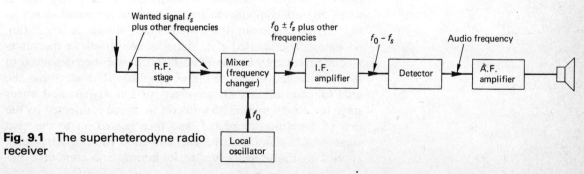

Fig. 9.1 The superheterodyne radio receiver

the radio-frequency stage to the mixer (or frequency changer). The r.f. stage is not provided to select the wanted signal but chiefly to prevent certain particularly troublesome frequencies reaching the mixer stage. In the mixer stage the input frequencies are combined with the output of the local oscillator, at a frequency f_o, to generate components at a large number of new frequencies. Amongst the newly generated frequencies are components at the sum and the difference of the wanted signal and the local oscillator frequencies, i.e. at $f_o \pm f_s$.

The *difference frequency* $f_o - f_s$ is known as the INTERMEDIATE FREQUENCY and is selected by the intermediate frequency (i.f.) amplifier. The intermediate frequency is a fixed frequency and this means that, when a receiver is tuned to receive a signal at a particular frequency, the local oscillator frequency is adjusted so that the correct difference frequency is obtained. The amplified output of the i.f. amplifier is applied to the detector stage and it is here that the information contained in the modulated signal is recovered. The detected signal is amplified to the required power level by the audio-frequency amplifier and is then fed to the loudspeaker, telephone or other output device.

A number of differences exist between receivers designed for the reception of amplitude- and frequency-modulated transmissions and i.s.b./s.s.b. radio-telephony signals. The main differences are as follows:

(a) The r.f. stage in an a.m. broadcast receiver may not include amplification whereas the other types of receiver always provide gain.

(b) The bandwidths of the r.f. and the i.f. stages are considerably different; often h.f. communication receivers have a variable bandwidth facility since they may be designed to cater for a number of different kinds of signal.

(c) Mainly because of the different bandwidths required, different intermediate frequencies are used.

(d) Different types of detector circuit are used.

Most frequency modulation broadcast receivers are also capable of the reception of amplitude-modulation signals; when discrete components are used the arrangement shown in Fig. 9.2 is common; the switches are shown in their f.m. positions. The wanted f.m. signal is converted to the intermediate frequency by the F.M. TUNER and then delivered to the first common stage of the i.f. amplifier. This stage has the dual function of first i.f. amplifier for f.m. signals and mixer stage for a.m. signals. The wanted f.m. signal is selected by the first i.f. amplifier, amplified, and then passed on to the next stage of i.f. amplification. The amplified f.m. signal is then applied to the detector where its information content is ex-

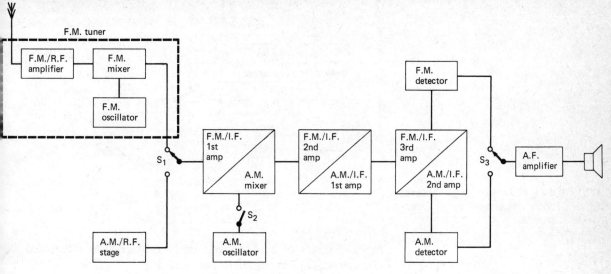

Fig. 9.2 F.M./A.M. superheterodyne radio receiver

tracted and then passed to the a.f. amplifier. When amplitude modulation signals are to be received, all the switches shown are operated and the first i.f. stage then acts as the a.m. mixer. The amplitude-modulated i.f. signal is selected by the second i.f. amplifier stage, which now acts as the first i.f. amplifier, and is then applied to the a.m. detector. The use of dual function stages is common since it results in a considerable reduction in the number of components needed.

Integrated circuits are increasingly employed in radio receivers, and Figs. 9.3 and 9.4 show two examples of modern practice. Fig. 9.3 shows the block diagram of an a.m. receiver; one i.c. performs the functions of the mixer, the i.f. amplifier, the detector, and the audio pre-amplifier; the other i.c. acts as the a.f. power amplifier. Provided externally to these integrated circuits are the components forming the r.f. stage and all the necessary inductors, capacitors and resistors for the other stages which cannot be formed within the i.c. package. The

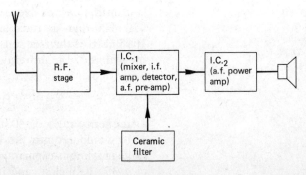

Fig. 9.3 Superheterodyne radio receiver using integrated circuits

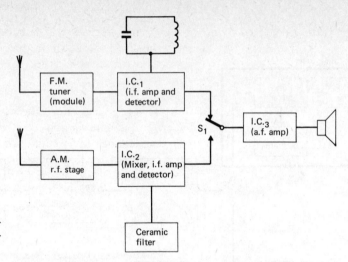

Fig. 9.4 F.M./A.M. superheterodyne radio receiver using integrated circuits.

block diagram of an f.m./a.m. receiver which uses integrated circuits is given in Fig. 9.4. The a.m. and f.m. sections of the receiver are completely separate up to the outputs of the two detector stages. The f.m. signal is amplified and frequency changed by the (non-integrated) f.m. tuner (often in module form), and is then passed on to an i.c. which performs the functions of both the i.f. amplifier and the f.m. detector; the selectivity of the i.f. amplifier is determined by an external inductor/capacitor network. The a.m. signal is received by a normal r.f. stage and is then fed to an integrated circuit which acts as the mixer, the i.f. amplifier and the a.m. detector. The audio-frequency outputs of the two detectors are connected, via a switch, to the common audio-frequency amplifier. The selectivity of the a.m. i.f. amplifier is determined by a *ceramic* filter.

Choice of Local Oscillator Frequency

The intermediate frequency of a superheterodyne radio receiver is the difference between the wanted signal frequency and the local oscillator frequency. Two possibilities exist: the local oscillator frequency can be higher than the signal frequency, or vice versa.

Consider a receiver with an intermediate frequency of 470 kHz that is tunable over the band from 525 kHz to 1605 kHz. If the frequency of the local oscillator is higher than the wanted signal frequency the oscillator must be tunable from

$$(525 + 470) = 995 \text{ kHz to } (1605 + 470) = 2075 \text{ kHz}$$

a frequency ratio of 2075/995, or $2 \cdot 085 : 1$. Such a frequency ratio would require the use of a variable capacitor having a ratio maximum-capacitance/minimum-capacitance of $(2.085)^2$, or $4 \cdot 35 : 1$. Such a capacitance ratio is easily obtained.

The alternative is to make the signal frequency higher than the local oscillator frequency. The oscillator frequency must then be variable from

$$(525 - 470) = 55 \text{ kHz to } (1605 - 470) = 1135 \text{ kHz}$$

This is a frequency ratio of 1135/55, or 20·64 : 1 and requires a capacitance ratio of $(20·64)^2$, or 425·9 : 1. Such a large capacitance ratio could not be obtained with a single variable capacitor and so tuning would not be as easy or cheap to achieve.

It is therefore usual to make the local oscillator frequency higher than the wanted signal frequency, i.e.

$$f_0 = f_s + f_i \tag{9.1}$$

The sum frequency component of the mixer output is not chosen for the intermediate frequency because it would mean that the latter would have to be greater than the highest frequency in the tuning range of the receiver. The various factors leading to the choice of intermediate frequency will be discussed later; here it will suffice to say that use of the sum frequency would prevent the use of the optimum intermediate frequency.

Image Channel Interference

No matter what frequency a superheterodyne receiver is tuned to, there is always another frequency that will also produce the intermediate frequency. This other frequency is known as the IMAGE FREQUENCY. The image signal has a frequency f_{im} such that the difference between it and the local oscillator frequency is equal to the intermediate frequency, f_i, i.e.

$$f_i = f_{im} - f_0$$

Substituting for f_0 from equation (9.1),

$$f_i = f_{im} - (f_s + f_i)$$

or

$$f_{im} = f_s + 2f_i \tag{9.2}$$

The image signal is thus separated from the wanted signal by twice the intermediate frequency. The image signal must be prevented from reaching the mixer or it will produce an interference signal which, since it is at the intermediate frequency, cannot be eliminated by the selectivity of the i.f. amplifier. The r.f. stage must include a resonant circuit with sufficient selectivity to reject the image signal when tuned to the wanted signal frequency. Tuning is necessary because the wanted signal frequency, and hence the image signal frequency, will vary. It is not difficult to obtain a resonant circuit with good enough selectivity to accept the wanted signal and reject the image signal when their separation is an appreciable fraction of the wanted signal frequency. As the signal frequency is increased, the fractional frequency separation becomes smaller and the image rejection less efficient.

Any vestige of the image signal reaching the mixer will produce a signal appearing as crosstalk at the output of the receiver. If a signal at a few kilohertz away from the image signal should reach the mixer, the two i.f. signals produced would beat together to produce a whistle at the output of the receiver.

The *image response ratio* is the ratio, in decibels, of the voltages at the wanted signal and image signal frequencies necessary at the receiver input terminals to produce the same audio output.

EXAMPLE 9.1

(**A**) A superheterodyne radio receiver has an intermediate frequency of 470 kHz and is tuned to 1065 kHz. Calculate (a) the frequency of the local oscillator, and (b) the frequency of the image signal.

Solution
From equation (9.1)

$$f_0 = 1065 + 470 = 1535 \text{ kHz} \qquad (Ans.)$$

and from equation (9.2)

$$f_{im} = 1065 + 940 = 2005 \text{ kHz} \qquad (Ans.)$$

(**B**) A superheterodyne radio receiver has an intermediate frequency of 10.7 MHz and is tuned to 97.3 MHz. Calculate (a) the frequency of the local oscillator and (b) the image channel frequency.

Solution
From equation (9.1)

$$f_0 = 97.3 + 10.7 = 108.0 \text{ MHz} \qquad (Ans.)$$

and from equation (9.2)

$$f_{im} = 97.3 + 21.4 = 118.7 \text{ MHz} \qquad (Ans.)$$

I.F. Breakthrough

If a signal at the intermediate frequency is picked up by an aerial and allowed to reach the mixer, it will reach the i.f. amplifier and interfere with the wanted signal. Such a signal must therefore be suppressed in the r.f. stage by an *i. f. trap*. The i.f. trap consists of either a parallel-resonant circuit, tuned to the intermediate frequency, connected in series with the aerial lead, or a series-resonant circuit, also tuned to the intermediate frequency, connected between the aerial lead and earth. In the first circuit the i.f. trap has a high impedance and blocks the passage of the unwanted i.f. signal; in the second circuit the i.f. trap has a low impedance and shunts the unwanted signal to earth.

EXAMPLE 9.2

A superheterodyne radio receiver has an intermediate frequency of 465 kHz and is tuned to receive an unmodulated carrier at 1200 kHz. Calculate the frequency of the audio output signal if present at the mixer input there are also (*a*) a 1208 kHz, and (*b*) a 462 kHz sinusoidal signal.

Solution

(*a*) The local oscillator frequency is $465 + 1200 = 1665$ kHz, and hence the 1208 kHz signal produces a difference frequency output from the mixer of $1665 - 1208 = 457$ kHz. If the i.f. bandwidth is only 9 kHz centred on 465 kHz, the 457 kHz signal will be rejected.

(*b*) The 462 kHz signal will appear at the mixer output and will be passed by the i.f. amplifier and will beat with the 465 kHz signal to produce a 3 kHz tone at the receiver output.

Other Sources of Interference

A superheterodyne receiver is also exposed to a number of other sources of interference. *Co-channel interference* is due to another signal at the same frequency and cannot be eliminated by the receiver itself. When it occurs it is the result of unusual propagation conditions making it possible for transmissions from a distant (geographically) station to be picked up by the aerial. Harmonics of the local oscillator frequency may combine with unwanted stations or with harmonics produced by the mixer to produce various difference frequency components, some of which may fall within the passband of the i.f. amplifier. It is also possible for two r.f. signals arriving at the input to the mixer to beat together and produce a component at the intermediate frequency.

The transfer and mutual characteristics of a bipolar or a field-effect transistor exhibit some non-linearity and as a result the output waveform will contain components at frequencies which were not present at the input. If, for example, the input signal contains components at frequencies f_1 and f_2, the output may contain components at frequencies $f_1 \pm f_2$, $2f_1 \pm f_2$, $2f_2 \pm f_1$, etc. These new frequencies are known as *intermodulation products*. Intermodulation can take place in both the r.f. amplifier and the mixer if the input signal level is so high that the active device is operated non-linearly. If two unwanted strong signals, separated in frequency by the intermediate frequency, or near to it, are present at the r.f. amplifier or mixer stages, they will produce an interfering component that will not be rejected by the i.f. amplifier.

One example of intermodulation which particularly affects v.h.f./f.m. receivers is known as *half i.f. interference*. Consider two signals at frequencies f and $f + \frac{1}{2}f_{if}$ to be present at the r.f. stage and to produce a voltage at their difference frequency. The second harmonic of this component is $2[f - (f + \frac{1}{2}f_{if})]$ which is equal to the intermediate frequency of the receiver.

Intermodulation interference can be reduced by operating the r.f. stage as linearly as possible and if possible rejecting one of the input voltages generating the interference.

Local Oscillator Radiation

The local oscillator operates at a radio frequency and may well radiate either directly or by coupling to the aerial. Direct radiation is limited by screening the oscillator. Radiation from the aerial is reduced by using an r.f. amplifier to prevent the oscillator voltage reaching the aerial. Radiation of the local oscillator frequency does not have a detrimental effect on the receiver in which it originates but is a source of interference to other nearby receivers.

Cross-modulation

Cross-modulation is the transfer of the amplitude modulation of an unwanted carrier onto the wanted carrier and is always the result of non-linearity in the mutual characteristic of the r.f. amplifier or of the mixer. If the amplitude of the input signal is small, or the mutual characteristic is essentially square law, cross-modulation will not occur. The unwanted signal may lie well outside the passband of the i.f. amplifier but, once cross-modulation has occurred, it is not possible to remove the unwanted modulation from the wanted carrier.

Cross-modulation is only present as long as the unwanted carrier producing the effect exists at the aerial, and it can be minimized by linear operation of the r.f. stage and by increasing the selectivity of the r.f. stage to reduce the number of large-amplitude signals entering the receiver. It is also helpful to avoid applying a.g.c. to the r.f. stage and, if large amplitude signals are expected, to use a switchable aerial attenuator to reduce the signal level and avoid overload with its consequent non-linearity. Cross-modulation does not occur in a frequency-modulation receiver because the unwanted amplitude variations will be removed by the limiter stage.

Blocking

Blocking is an effect in which the gain of one or more stages in a radio receiver is reduced by an interfering signal of sufficient strength to overload the stage, or to excessively operate the a.g.c. system of the receiver. The practical result of blocking is that the wanted signal output level falls every time the interfering signal is received.

Selectivity

The SELECTIVITY *of a radio receiver is its ability to discriminate between the wanted signal and all the other signals picked up by the aerial,* particularly the adjacent-channel signals. The selectivity of a receiver is usually quoted by means of a graph showing the output of the receiver, in dB relative to the maximum output, plotted against the number of kHz off-tune or by quoting some points on this graph. For example, the selectivity of an h.f. receiver may be quoted as −6 dB at 3 kHz bandwidth and −60 dB at 12 kHz bandwidth. Fig. 9.5 shows typical selectivity curves for a.m. broadcast, f.m. broadcast, and h.f. s.s.b. communication receivers. Clearly, there are large differences between the 3 dB bandwidths of the three receivers; the a.m. broadcast receiver has a 3 dB bandwidth of

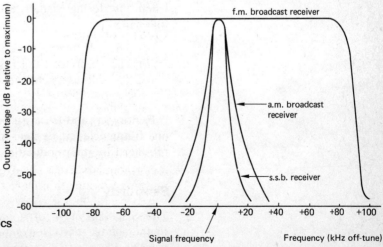

Fig. 9.5 Selectivity characteristics of radio receivers

about 9 kHz, the s.s.b. receiver approximately 3 kHz, but the f.m. broadcast receiver's bandwidth is about 200 kHz.

The adjacent channel selectivity of a radio receiver is mainly determined by the gain/frequency characteristic of the i.f. amplifier.

The ADJACENT CHANNEL RATIO is the ratio, in dB, of the input voltages at the wanted and the adjacent channel frequencies necessary for the adjacent channel to produce an output power 30 dB smaller than the signal power.

EXAMPLE 9.3

A superheterodyne radio receiver is tuned to a certain frequency at which an input signal of 15 μV produces an output of 50 mW. If the input voltage at the adjacent-channel frequency needed to produce -30 dB output power is 1.5 mV calculate the adjacent channel ratio.

Solution

$$\text{Adjacent channel ratio} = 20 \log_{10}\left(\frac{1.5\times10^{-3}}{15\times10^{-6}}\right) = 40\text{ dB} \qquad (Ans.)$$

The 6 dB and the 60 dB bandwidths are often known, respectively, as the nose and the skirt bandwidths. The NOSE BANDWIDTH is the range of frequencies over which a signal can be received with little practical loss of strength. The SKIRT BANDWIDTH is the band of frequencies over which it is possible to receive a strong signal. The ratio of the skirt bandwidth to the nose bandwidth is known as the *shape factor.* Thus the h.f. s.s.b. receiver quoted earlier has a nose band-width of 3 kHz, a skirt bandwidth of 12 kHz, and a shape factor of 12 kHz/3 kHz or 4.0.

The selectivity curves shown in Fig. 9.5 relate to a single input frequency and do not entirely predict the performance of a receiver when signals at several different frequencies are simultaneously received. The effective selectivity of a receiver when interfering signals are present is determined by the following:

(*a*) The selectivity provided by the r.f. stage.
(*b*) The ability of the r.f. stage to handle strong signals.
(*c*) The adjacent-channel selectivity provided by the i.f. stage.

Factors (*a*) and (*b*) arise because of the possibility of spurious frequencies at or near the intermediate frequency being produced by intermodulation.

Sensitivity

The SENSITIVITY *of a radio receiver is the smallest input signal voltage that is required to produce a specified output power with a specified signal-to-noise ratio.* For amplitude-modulation receivers, the specified output power is usually 50 mW with a signal-to-noise ratio of 20 dB and the input signal modulated 30% at 1000 Hz (or 400 Hz). For an f.m. receiver a signal-to-noise ratio of 40 dB is required with the input signal modulated by a 1000 Hz signal to give 30% modulation. (This means that the frequency deviation produced should be 30% of the rated system deviation, i.e. for the v.h.f. sound broadcast system 30% of 75 kHz is 22.5 kHz.)

It is necessary to include signal-to-noise ratio in the definition of sensitivity, otherwise the output power could consist mainly of noise and be of little use.

The sensitivity of a radio receiver is determined by

(*a*) The overall voltage gain of its individual stages.
(*b*) The gain/frequency characteristic of the r.f. stage.
(*c*) The noise generated by thermal agitation in its input stages.

This means that the sensitivity is directly related to the *noise figure* of the receiver.

Typical figures for sensitivity are (*a*) a.m. broadcast receiver 50 μV, (*b*) f.m. broadcast receiver 2 μV, and (*c*) s.s.b. receiver 1 μV.

Noise Figure

The output of a radio receiver must always contain some noise and the receiver must be designed so that the output signal-to-noise ratio is always at least as good as the minimum figure required for the system. The noise appearing at the receiver's output terminals originates from two sources; noise picked up by the aerial and noise generated within the receiver [EIII]. Because of the internally generated noise, the signal-to-noise ratio at the output terminals is *always* less than the input signal-to-noise ratio. The *noise figure or factor* of a radio receiver is a measure of the degree to which the receiver degrades the input signal-to-noise ratio. The NOISE FACTOR F is related to the input and output signal-to-noise ratios by equation (9.3):

$$F = \frac{\text{Input signal-to-noise ratio}}{\text{Output signal-to-noise ratio}} \qquad (9.3)$$

An ideal receiver would have no internal noise sources and would not degrade the input signal-to-noise ratio; hence its noise figure would be unity or zero dB.

EXAMPLE 9.4

The signal-to-noise ratio at the input to a communication receiver is 40 dB. If the receiver has a noise figure of 12 dB calculate the output signal-to-noise ratio.

Solution
From equation (9.3)

$$\text{Output signal-to-noise ratio} = \frac{\text{Input signal-to-noise ratio}}{\text{Noise factor}}$$

or in dB

$$\text{Output signal-to-noise ratio} = 40 - 12 = 28 \text{ dB} \qquad (Ans.)$$

The Radio-frequency Stage

The radio-frequency stage of a superheterodyne radio receiver must perform the following functions:

(*a*) It must couple the aerial to the receiver in an efficient manner.
(*b*) It must suppress signals at or near the image and the intermediate frequencies.
(*c*) At frequencies in excess of about 3 MHz it must provide gain.
(*d*) It must operate linearly to avoid the production of cross-modulation.
(*e*) It should have sufficient selectivity to minimize the number of frequencies appearing at the input to the mixer that could result in intermodulation products lying within the passband of the i.f. amplifier.

At frequencies up to about 3 MHz or so, the noise picked up by an aerial is larger than the noise generated within the receiver. An r.f. amplifier will amplify the aerial noise as well as the signal and produce little, if any, improvement in the output signal-to-noise ratio. At higher frequencies the noise picked up by the aerial falls and the constant-level receiver noise becomes predominant; the use of r.f. gain will then improve the output signal-to-noise ratio. An r.f. amplifier also permits the use of two or more tuned circuits in cascade, with a consequent improvement in the image response ratio.

The Mixer Stage

The function of the mixer stage is to convert the wanted signal frequency into the intermediate frequency of the receiver. This process is carried by *mixing* the signal frequency with the output of the local oscillator and selecting the resultant difference frequency.

The local oscillator must be capable of tuning to any frequency in the band to which the receiver is tuned *plus* the intermediate frequency, i.e. $f_0 = f_s + f_{if}$. The ability of a receiver to remain tuned to a particular frequency without drifting depends upon the frequency stability of its local oscillator. In an a.m. broadcast receiver the demands made on the oscillator in terms of frequency stability are not stringent since the receiver is tuned by ear. High-frequency communications receivers need greater frequency stability mainly because the channel bandwidth is narrow. Receivers operating at one or more fixed frequencies can use a crystal oscillator, frequency changes involving crystal switching. When a receiver is to be tunable over a band of frequencies an *L-C* oscillator with *automatic frequency control* or a *frequency synthesizer* must be used.

The frequency stability of an i.s.b./s.s.b. receiver should be good enough to ensure that the tuning of the receiver will not drift from its nominal value by more than about 20 Hz over a long period of time. This is necessary because any change in the local oscillator frequency will cause a corresponding shift in the frequency of the output signal. If, for example, the oscillator frequency should be 10 Hz too high, then all the components of the output signal will also be 10 Hz too high. If data and/or v.f. telegraph signals are to be received, the maximum permissible frequency drift is only ± 1 Hz. Generally, the long-term frequency stability of an h.f. communications receiver is better than 1 part in 10^7.

Ganging and Tracking

When a superheterodyne radio receiver is tuned to receive a particular signal frequency, the resonant circuit(s) in the r.f. stage must be tuned to that frequency and the tuned circuit of the local oscillator must be tuned to a frequency equal to the sum of the signal and the intermediate frequencies. Clearly it is convenient if the tuning of these circuits can be carried out by a single external control. To make this possible the tuning capacitors are mounted on a common spindle so that they can be simultaneously adjusted; this practice is known as GANGING. The maintenance of the correct frequency difference (the intermediate frequency) between the r.f. stage and local oscillator frequencies is known as TRACKING.

It is possible to achieve nearly perfect tracking over one particular waveband if the plates of the oscillator tuning capacitor are carefully shaped, but this practice requires a different capacitor for each waveband and involves design problems. Most radio receivers use identical tuning capacitors for the r.f. and oscillator circuits and modify the capacitance values by means of *trimmer* and/or *padder* capacitors.

Consider a receiver designed to tune over the medium frequency band of 525–1605 kHz and have an intermediate frequency of 470 kHz. Suppose that the (identical) variable tuning capacitors used in the r.f. and oscillator circuits have a capacitance range of (maximum capacitance − minimum capacitance) of 400 pF. The r.f. stage must tune over the band 525–1605 kHz; this is a frequency ratio of 1605/525 or 3.057 : 1 and requires the tuning capacitance to provide a capacitance ratio of $3.057^2 : 1$ or 9.346 : 1. Therefore, if the minimum capacitance needed is denoted by x, then

$$9.346x = x + 400$$

$$x = 400/8.346 = 47.93 \text{ pF} \simeq 48 \text{ pF}$$

The maximum capacitance must then be $48 + 400 = 448$ pF. If the minimum capacitance of the variable capacitor plus the inevitable stray capacitances is not equal to 48 pF, a TRIMMER capacitance must be connected in parallel with the tuning capacitance. For example, if the minimum tuning capacitance plus stray capacitances adds up to 40 pF, an 8 pF trimmer will be required (see Fig. 9.6). The inductance L_1 required to tune the r.f. stage to the wanted signal frequency can be calculated using the expression $f_0 = 1/2\pi\sqrt{(LC)}$, remembering that the minimum capacitance corresponds to the maximum frequency and vice versa. Thus

$$L_1 = 1/(4\pi^2 \times 1605^2 \times 10^6 \times 48 \times 10^{-12}) \simeq 205 \ \mu\text{H}$$

The oscillator must be able to tune over the frequency band

$$(525 + 470) = 995 \text{ kHz to } (1605 + 470) = 2075 \text{ kHz}$$

This is a frequency ratio of 2075/995 or 2.085 : 1 and requires a capacitance ratio of 2.085^2 or 4.349 : 1. This capacitance ratio must be obtained using the same tuning capacitor as before and so either the minimum capacitance must be increased or the maximum capacitance must be decreased.

Use of a Trimmer Capacitor

Let the minimum capacitance needed in the local oscillator be x pF then

$$4.349x = x + 400$$

$$x = 400/3.349 = 119.4 \text{ pF} \simeq 119 \text{ pF}$$

The minimum capacitance of the oscillator tuning circuit can be increased to this value by connecting a trimmer capacitor in parallel with the variable capacitor. Assuming the minimum capacitance of the variable plus strays to be the same as in the r.f. circuit, i.e. 40 pF, then a trimmer capacitor of $119 - 40 = 79$ pF is needed (see Fig. 9.7). The tuning inductance L_2 is

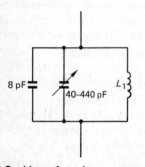

Fig. 9.6 Use of a trimmer capacitor in the r.f. stage

8 pF

40–440 pF

L_1

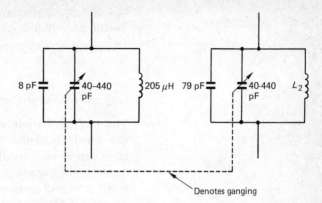

Fig. 9.7 Use of a trimmer capacitor
in the oscillator

Denotes ganging

easily calculated using $L_2 = 1/4\pi^2 f^2_{0(max)} C_{min}$ and is equal to 49.4 μH. Unfortunately, correct tracking will not be maintained between the r.f. and local oscillator circuits as the receiver is tuned over its frequency bands.

EXAMPLE 9.5

A superheterodyne radio receiver employs ganged capacitors in its aerial and local oscillator circuits with an additional parallel capacitor in the local oscillator circuit. As the capacitance in the signal circuit varies from 80 pF to 320 pF the receiver is tuned from 1200 kHz to 600 kHz. If the local oscillator capacitance variation is from 160 pF to 400 pF, and the intermediate frequency is 433 kHz, what is (i) the frequency to which the receiver is tuned when the signal circuit capacitance is 200 pF, (ii) the local oscillator frequency when the local oscillator capacitance is 280 pF, (iii) the tracking error when the capacitance is at the mid-point of its range?   (C & G)

Solution
For both the r.f. and the local oscillator circuits the maximum frequency corresponds to the minimum capacitance. Therefore,

(i) $1200 \times 10^3 = 1/2\pi\sqrt{(L \times 80 \times 10^{-12})}$ (9.4)

$$f = 1/2\pi\sqrt{(L \times 200 \times 10^{-12})}$$ (9.5)

Dividing equation (9.4) by equation (9.5),

$$\frac{1200 \times 10^3}{f} = \sqrt{\frac{200}{80}}$$

$f = 1200 \times 10^3/1.581 = 758.95$ kHz   (*Ans.*)

(ii) Maximum oscillator frequency $= 1200 + 433 = 1633$ kHz

$$1633 \times 10^3 = 1/2\pi\sqrt{(L \times 160 \times 10^{-12})}$$

$$f = 1/2\pi\sqrt{(L \times 280 \times 10^{-12})}$$

$$\frac{1633 \times 10^3}{f} = \sqrt{\frac{280}{160}} = 1.323$$

$f = 1633 \times 10^3/1.323 = 1234.43$ kHz   (*Ans.*)

(iii) The midpoint of the capacitance range corresponds to the values used in parts (i) and (ii). Therefore,

$$\text{Tracking error} = 1234.43 \text{ kHz} - (758.95 + 433) \text{ kHz}$$
$$= 42.48 \text{ kHz} \quad (Ans.)$$

Use of a Padder Capacitor

The alternative method of reducing the capacitance ratio in the local oscillator circuit is the connection of a *padder* capacitor in series with the tuning capacitor. Suppose that the same tuning capacitor and frequencies as before are used. The minimum and maximum capacitances are then 40 pF and 440 pF. If the padder capacitor is denoted by C_p then, since the required capacitance ratio is 4.349 : 1,

$$\frac{440C_p}{440+C_p} = 4.349 \times \frac{40C_p}{40+C_p}$$

$$440 \times 40 + 440C_p = 4.349 \times 40 \times 440 + 4.349 \times 40C_p$$

$$C_p(440 - 4.349 \times 40) = 40 \times 440(4.349 - 1)$$

$$C_p = \frac{40 \times 440 \times 3.349}{440 - 4.349 \times 40} = 221.6 \text{ pF} \approx 222 \text{ pF}$$

Fig. 9.8 shows the arrangement of the r.f. and local oscillator circuits.

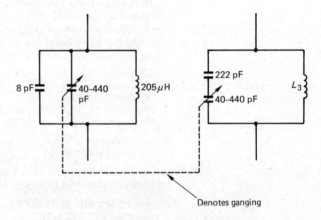

Fig. 9.8 Use of a padder capacitor

The total capacitance of the circuit will now vary from

$$\frac{40 \times 222}{40 + 222} \text{ or } 33.89 \text{ pF} \quad \text{to} \quad \frac{440 \times 222}{440 + 222} \text{ or } 147.55 \text{ pF}$$

The inductance L_3 required to tune the circuit can be calculated using the maximum frequency of 2075 kHz and the minimum capacitance of 33.89 pF (or vice versa). Thus,

$$L_3 = 1/(4\pi^2 \times 2075^2 \times 10^6 \times 33.89 \times 10^{-12}) = 173.6 \text{ } \mu\text{H}$$
$$\simeq 174 \text{ } \mu\text{H}$$

Again, correct tracking is not obtained over most of the tuning range of the receiver.

The tracking error does not have much effect on the quality of the audio output signal since the tuning of the receiver positions the wanted signal into the middle of the passband of the i.f. amplifier. This means that the intermediate frequency is correct and the tracking error exists in the r.f. stage. The error in tuning the r.f. stage has little, if any, effect upon the adjacent channel selectivity but both the sensitivity and the image channel rejection are worsened. Tracking is not a problem in v.h.f. receivers because the required frequency and capacitance ratios are small.

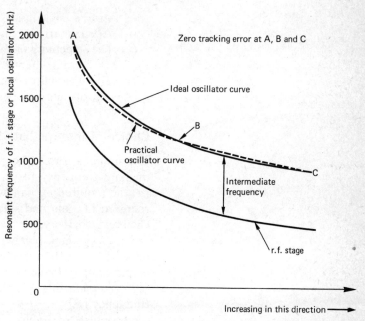

Fig. 9.9 Tracking curves

If both a trimmer and a padder capacitor are used, *three-point tracking* can be obtained and the tracking error reduced to a small figure. Three-point tracking is illustrated by the curves of Fig. 9.9: the r.f. circuit and the *ideal* oscillator curves are always separated by a frequency difference equal to the intermediate frequency of the receiver. The practical curve, shown dotted, has zero tracking error at the three points marked A, B and C; elsewhere the error is small.

The Intermediate Frequency Amplifier

The purpose of the i.f. amplifier in a superheterodyne radio receiver is to provide most of the gain and the selectivity of the receiver. Most broadcast receivers utilize the impedance/frequency characteristics of single- or double-tuned circuits to obtain the required selectivity, but many receivers use *ceramic filters*, particularly when an integrated circuit is used as the i.f. amplifier. Narrow-band communication receivers must possess very good selectivity and very often employ one or more *crystal filters* to obtain the necessary gain/frequency response. The use of a ceramic or a crystal filter to provide the selectivity of a radio receiver offers a number of advantages over the use of *L-C* networks:

(a) A very narrow bandwidth can be obtained.
(b) The selectivity of the receiver does not depend upon the correct alignment of the i.f. amplifier.
(c) The selectivity of the filter is not affected by the application of automatic gain control to the receiver.

Choice of Intermediate Frequency

The main factors to be considered when choosing the intermediate frequency for a superheterodyne radio receiver are (a) the required i.f. bandwidth, (b) interference signals, (c) the required i.f. gain and stability, and (d) the required adjacent channel selectivity.

The minimum bandwidth demanded of an i.f. amplifier depends upon the type of receiver and is about 9 kHz for an amplitude-modulation broadcast receiver. Since the bandwidth of a coupled-tuned circuit is proportional to its resonant frequency ($B = \sqrt{2}f_0/Q$) the larger the bandwidth required the higher must be the intermediate frequency. The intermediate frequency should not lie within the tuning range of the receiver, so that the r.f. stage can include an i.f. trap to prevent i.f. interference. However, to simplify the design and construction of the i.f. amplifier, the intermediate frequency should be as low as possible. Adequate adjacent channel selectivity is easier to obtain using a low intermediate frequency, but on the other hand, image channel rejection is easier if a high intermediate frequency is selected.

The intermediate frequency chosen for a receiver must be a compromise between these conflicting factors. Most amplitude-modulated broadcast receivers employ an intermediate frequency of between 450 and 470 kHz; but frequency-modulation broadcast receivers, which require an i.f. bandwidth of about 200 kHz, use an intermediate frequency of 10.7 MHz.

The Detector Stage and Automatic Gain Control

The function of the detector stage in a radio receiver is to recover the information modulated onto the carrier wave appearing at the output of the i.f. amplifier. Most a.m. broadcast receivers use the *diode detector* because of its simplicity and good performance but i.c. versions often use the transistor detector. The transistor detector is not often used in discrete form for broadcast receivers because of its limited dynamic range, but it is used in v.h.f. communication receivers where its ability to provide an amplified a.g.c. voltage and its gain are an advantage. Most f.m. broadcast receivers use the ratio detector but high-quality broadcast receivers may use the Foster–Seely circuit; when the latter circuit is used the detector must be preceded by a limiter stage. When the detector stage is part of an integrated circuit, the quadrature detector or, less often, the phase-locked loop detector is used. High-frequency i.s.b./s.s.b. communication receivers generally use some form of balanced or product demodulator.

Automatic Gain Control

The field strength of the wanted signal at the aerial is not constant but fluctuates widely because of changes in propagation conditions. Automatic gain control (a.g.c.) is applied to a radio receiver to maintain the carrier level at the input to the detector at a more or less constant value even though the level at the aerial may vary considerably. A.G.C. ensures that the audio output of the receiver varies only as a function of the modulation of the carrier and not with the carrier level itself. The use of a.g.c. also ensures that a large receiver gain can be made available for the reception of weak signals without causing overloading of the r.f. amplifier stages, with consequent distortion, by strong signals. Further, a reasonably constant output level is obtained as the receiver is tuned from one station to another.

In an f.m. receiver automatic gain control is often fitted to ensure (*a*) that the signal arriving at the input terminals of the limiter is large enough for the limiting action to take place and (*b*) that overloading of the r.f. and i.f. amplifier stages does not occur. In some cases the automatic gain control of an f.m. receiver may mean switching into the r.f. stage of one or more stages of an attenuator.

The basic idea of an a.g.c. system is illustrated by Fig. 9.10; a direct voltage is developed, either in the detector stage or in the amplitude limiter, that is proportional to the amplitude of the carrier signal appearing at the output of the i.f. amplifier. The gain of a transistor amplifier is a function of the d.c.

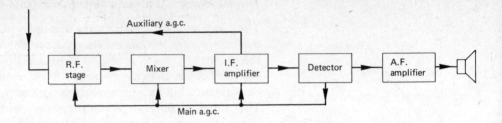

operating point of the transistor; hence if the a.g.c. voltage is applied to each of the controlled stages to vary their bias voltages, the gains of these stages will be under the control of the a.g.c. system. The polarity of the a.g.c. voltage should be chosen so that an increase in the carrier level, which will produce an increase in the a.g.c. voltage, will reduce the gain of each stage. This will, in turn, reduce the overall gain of the receiver and tend to restore the carrier level at the detector to its original value. Conversely, of course, if the carrier level should fall, the gain of the receiver will be increased to tend to keep the level at the detector very nearly constant. Another a.g.c. loop, known as *auxiliary a.g.c.*, is often provided to give extra control of the gain of the receiver and to limit the amplitude of strong input signals to prevent overloading of the r.f. amplifier and the consequent distortion and cross-modulation. In many f.m. receivers only auxiliary a.g.c. is fitted.

Main A.G.C.

Automatic gain control systems are either of the *simple* or the *delayed* type. In a SIMPLE A.G.C. SYSTEM the a.g.c. voltage is developed immediately a carrier voltage appears at the output of the i.f. amplifier. This means that the gain of the receiver is reduced below its maximum value when the wanted signal is weak and the full receiver gain is really wanted. This disadvantage of the simple a.g.c. system can be overcome by arranging that the a.g.c. voltage will not be developed until the carrier level at the detector has reached some pre-determined value—generally that at which the full audio-frequency power output can be developed. Such a system is known as a DELAYED A.G.C. SYSTEM. Fig. 9.11 shows, graphically, the difference between simple and delayed a.g.c. systems; in addition the performance of the ideal a.g.c. system is shown. It is evident that the ideal system is one in which no a.g.c. voltage is produced until the input voltage to the receiver exceeds some critical value and thereafter keeps the output level of the receiver perfectly constant. For economic reasons the majority of broadcast receivers use simple a.g.c.

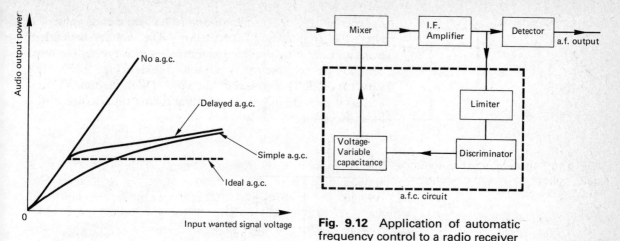

Fig. 9.11 Automatic gain control characteristics

Fig. 9.12 Application of automatic frequency control to a radio receiver

Automatic Frequency Control

The intermediate frequency bandwidth of a communication receiver operating in the u.h.f. band is only a small percentage of the carrier frequency. A relatively small percentage error in the frequency of the local oscillator may lead to the wanted signal being wholly or partly rejected by the selectivity of the i.f. amplifier. Some of the necessary frequency stability can, however, be obtained by a suitable choice of the type of oscillator to be used but the most stable types of oscillator cannot be tuned to different frequencies. The required frequency stability can be obtained by the use of AUTOMATIC FREQUENCY CONTROL (a.f.c.). To avoid distortion of the output signal caused by mistuning, many f.m. broadcast receivers also have a.f.c. fitted.

The basic principle of an a.f.c. system is shown by Fig. 9.12. The output of the i.f. amplifier is passed through an amplitude limiter and is then applied to the input terminals of a discriminator. The input circuit of the discriminator is tuned to the nominal intermediate frequency of the receiver and so the circuit produces an output voltage of zero whenever the intermediate frequency is correct. If, however, the intermediate frequency differs from its nominal value a direct voltage will appear at the output of the discriminator. The polarity of this direct voltage will depend upon whether the intermediate frequency is higher than, or lower than, its nominal value. Thus if a negative voltage is produced by an increase in frequency, then a fall in the intermediate frequency will result in a positive direct voltage at the output of the discriminator. The direct voltage is taken to a voltage-variable capacitance, the magnitude of which is a function of that voltage. The variable capacitance is a part of the frequency-determining network of the local oscillator and so a change in its value will alter the frequency of oscillation. The voltage-dependent capacitance can be provided in a number of ways but the most common is the use of a varactor diode (Fig. 9.13). The varactor diode D is connected in parallel with the tuned circuit and so it provides a part of the total tuning capacitance of the local oscillator.

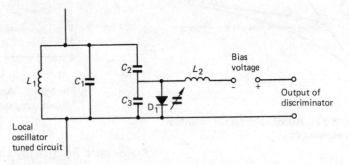

Fig. 9.13 Varactor diode control of local oscillator frequency

EXAMPLE 9.6

An a.f.c. system incorporates a discriminator with an output-voltage/input-frequency characteristic of 1 V/3 kHz and a voltage-controlled oscillator whose output-frequency/input-voltage characteristic is 15 kHz/V. Calculate the tuning error with the a.f.c. system operative if without the a.f.c. the frequency error would have been 24 kHz.

Solution
Let the final frequency error be f kHz. Then the d.c. output voltage of the discriminator is $f/3$ volts and this voltage will cause the frequency of the oscillator to be shifted by $(f/3) \times 15 = 5f$ kHz. The final tuning-error is equal to the original error minus the frequency correction provided by the a.f.c. system. Therefore,

$$f = 24 - 5f \quad \text{or} \quad f = 4 \text{ kHz} \quad (Ans.)$$

The PULL-IN or CAPTURE RANGE of an a.f.c. system is the maximum frequency error that can be reduced by the system. It is obviously necessary that an **a.f.c.** system is designed so that the capture range is larger than the maximum expected drift in the oscillator frequency. The HOLD-IN RANGE is the band of frequencies over which the controlled oscillator frequency can suddenly change without the control exerted by the a.f.c. system being lost.

Frequency Synthesis

Modern communication receivers often obtain the required degree of frequency stability by deriving the local oscillator frequency from a frequency synthesizer instead of using a.f.c.

The Audio-frequency Stage

The function of the audio-frequency stage of a radio receiver is to develop sufficient a.f. power to operate the loudspeaker or other receiving apparatus. The a.f. stage will include a volume control and sometimes treble and bass controls. The a.f. stage may also include a *squelch* or *muting* facility. A sensitive receiver will produce a considerable output noise level when there is no input signal because there will then be no a.g.c. voltage developed to limit the gain of the receiver. The noise unavoidably present at the input terminals of the receiver then receives maximum amplification. This noise output can cause considerable annoyance to the operator of the receiver and, to reduce or eliminate this annoyance, a SQUELCH circuit is fitted which disconnects, or severely attenuates, the gain of the a.f. amplifier whenever there is no input signal present.

The Double Superheterodyne Radio Receiver

To obtain good adjacent channel selectivity, the intermediate frequency of a superheterodyne radio receiver should be as low as possible, but to maximize the image channel rejection the intermediate frequency must be as high as possible. For receivers operating in the low and medium frequency bands it is possible to choose a reasonable compromise frequency. In the h.f. band it may prove difficult to select a suitable frequency and for this reason many receivers use two, or more rarely, three or four different intermediate frequencies.

The first intermediate frequency is chosen to give good image channel rejection ratio and the second frequency is chosen for good adjacent channel selectivity. Typically, the first intermediate frequency might be 3 MHz and the second intermediate frequency 100 kHz, although in modern receivers there is a tendency to use a very high first intermediate frequency, such as 35 MHz, to give a *very* good image rejection (usually the second intermediate frequency is then about 1 MHz).

The disadvantages of the double superheterodyne principle are the extra cost and complexity involved and the generation of extra spurious frequencies because there are two stages of mixing. The most serious of these new frequencies is the second/image channel frequency.

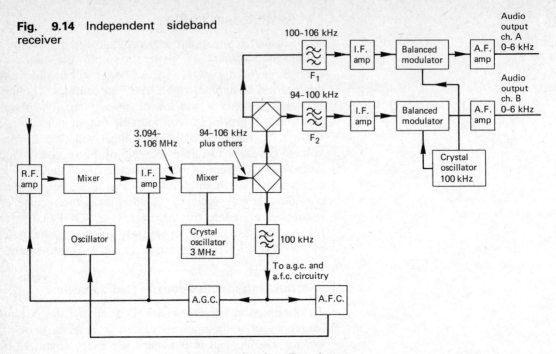

Fig. 9.14 Independent sideband receiver

Communication Receivers

Fig. 9.14 gives the block schematic diagram of an i.s.b. communication receiver. The received signal, in the 4–27.5 MHz frequency band, is first amplified and then frequency changed to the standard frequency band of 3.094–3.106 MHz. After further amplification, the signal is translated to the 94–106 kHz band. Channel filters, F_1 and F_2, select the signals appropriate to the channels, and the selected signals are then demodulated to obtain the original audio-frequency signal. A narrow passband filter tuned to 100 kHz selects the pilot carrier and applies it to the a.g.c. and a.f.c. circuitry of the receiver.

A more modern type of i.s.b. receiver now in use by the British Post Office is shown in block diagrammatic form in Fig. 9.15. It can be seen that this receiver uses four stages of mixing and i.f. amplification. The first intermediate frequency is above the tuning range of the receiver (3–27.5 MHz). Frequency synthesis is used to derive the first and the second/local oscillator frequencies, the synthesizer being controlled by a memory unit. The memory unit in conjunction with a digital circuit causes the appropriate band-pass filter to be switched into the aerial circuit to broadly select the wanted signal. Six filters are available covering, respectively, the frequency bands 3–4 MHz, 4–6 MHz, 6–9 MHz, 9–13 MHz, 13–19 MHz and 19–27.5 MHz. Selection of the wanted signal is achieved by the first, second and third i.f. amplifiers.

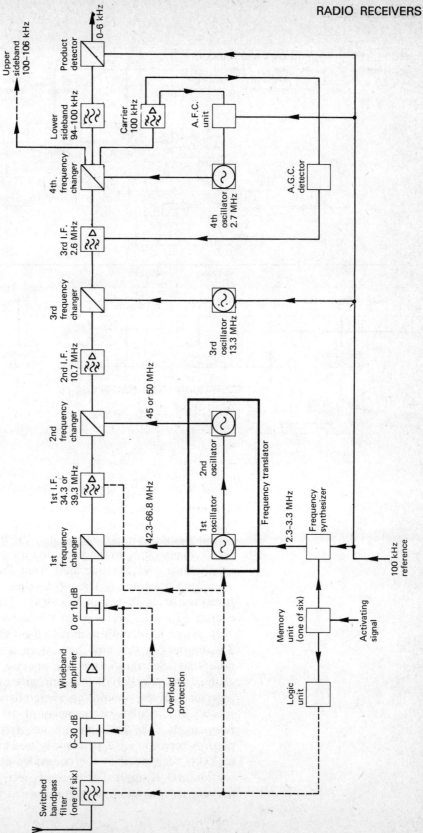

Fig. 9.15 A modern independent sideband receiver (From *Post Office Electrical Engineers' Journal*)

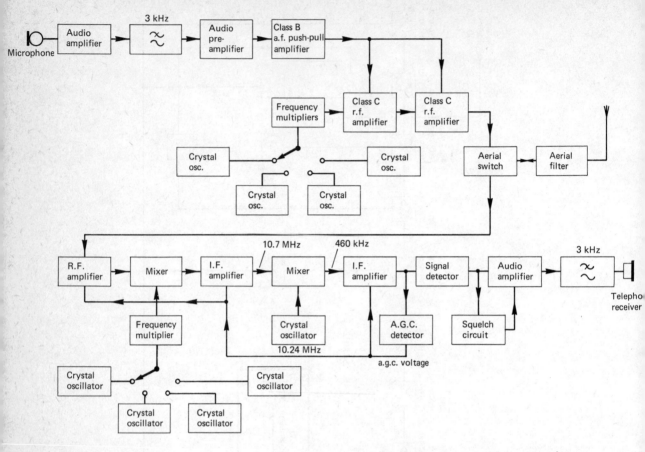

Fig. 9.16 A v.h.f. amplitude modulated transreceiver

Many amplitude- and frequency-modulated communication receivers operating in the v.h.f. and the u.h.f. bands are mobile installations and are operated from the same aerial as the associated transmitter. The block schematic diagram of a v.h.f *transreceiver* is shown in Fig. 9.16. The transmitter is only connected to the aerial when the *aerial switch* is depressed. This switch is very often mounted on the telephone handset. Transmitters of this type have been described in Chapter 8; note that the transmitter can operate on any one of four channels by selection of the appropriate crystal oscillator. The receiver is of the double superheterodyne type, using first and second intermediate frequencies of 10.7 MHz and 460 kHz respectively. The first intermediate frequency is fairly standard. A separate a.g.c. detector is used and this allows delayed a.g.c. to be applied to the controlled stages. Finally, squelch (or muting) is applied to the a.f. section of the receiver to prevent its operation when no signal is being received.

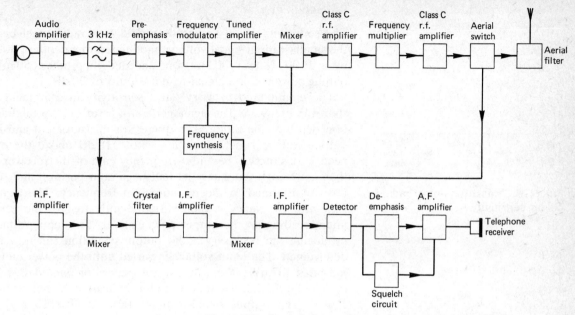

Fig. 9.17 A v.h.f. frequency modulated transreceiver

The block schematic diagram of a frequency-modulated transreceiver, which uses a frequency synthesizer to derive the wanted mixer frequencies, is shown in Fig. 9.17. Two stages of mixing and of i.f. amplification are provided, the selectivity of the second amplifier being provided by a crystal filter.

For both the transreceivers shown in Figs. 9.16 and 9.17 the choice of frequencies at various points in the circuit depends upon the frequency band in which the equipment is designed to operate. The various mobile bands are given elsewhere [TSII].

Measurement of Performance of Amplitude Modulation Receivers

A number of measurements can be carried out to determine the performance of a superheterodyne radio receiver. Some of these tests are appropriate for both amplitude- and frequency-modulated receivers while others only apply to one type of receiver. In this book only the more important of the amplitude modulation receiver measurements will be described; these are (a) sensitivity, (b) noise factor, (c) adjacent channel ratio, and (d) image channel response ratio.

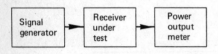

Fig. 9.18 Measurement of radio receiver sensitivity

(a) Sensitivity

The sensitivity of an amplitude modulation radio receiver is the smallest input signal voltage, modulated to a depth of 30% by a 1000 Hz (or 400 Hz) tone, needed to produce 50 mW output power with a signal-to-noise ratio of 20 dB.

The circuit used to carry out a sensitivity measurement is shown in Fig. 9.18. The signal generator is set to 30% modulation depth at the required frequency of measurement and its output voltage is set to a value about 10 dB above the expected sensitivity. The audio-frequency gain of the receiver is then set to approximately its half-maximum position and the receiver is tuned to the measurement frequency. The signal generator frequency is then varied slightly to give the maximum reading on the output power meter. The input voltage producing the necessary audio output condition can now be determined. The input voltage is varied until the power meter indicates 50 mW; then the signal generator *modulation* is switched off and the power meter indication is noted, say P mW. The output signal-to-noise ratio is now $10 \log_{10} \times (50/P)$ dB.

If this ratio is not equal to the required 20 dB the modulation of the signal generator is switched on again and the input voltage to the receiver is increased or decreased as appropriate. The a.f. gain is adjusted to obtain 50 mW indication on the power meter before the modulation is again switched off and the new signal-to-noise ratio determined. This procedure is repeated until the required output power of 50 mW is obtained *together with* 20 dB signal-to-noise ratio. The input signal voltage giving the required output conditions is the sensitivity of the receiver.

(b) Noise Factor

The noise factor F of a radio receiver is the ratio

$$F = \frac{\text{Noise power appearing at the output of the receiver}}{\text{That part of the above which is due to thermal agitation at the input terminals}} \quad (9.6)$$

This definition of noise factor is, for most conditions, equivalent to the previously quoted, equation (9.3), meaning of noise factor, i.e.

$$F = \frac{\text{Input signal-to-noise ratio}}{\text{Output signal-to-noise ratio}} \quad (9.3)$$

Fig. 9.19 shows the circuit used for the measurement of the noise factor of a receiver. With the noise generator switched off, the indication of the power output meter is noted. The

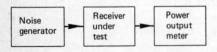

Fig. 9.19 Measurement of radio receiver noise figure

noise generator is then switched on and, without altering any of the receiver controls, its noise output is increased until the indication of the power meter is *exactly* double its previous value.

The noise output of the generator is directly proportional to the current indicated by an integral milliammeter and so the noise factor of the receiver is equal to

$$F = 20 I_a R \qquad (9.7)$$

where I_a is the indication of the milliammeter and R is the (matched) impedance of the receiver and the noise generator. If, as is often the case at v.h.f. and at u.h.f., $R = 50\ \Omega$, the noise factor of the receiver is equal to the milliammeter reading.

EXAMPLE 9.8

In a measurement of the noise factor of a $50\ \Omega$ input impedance radio receiver the reading of the output power meter is doubled when the noise generator's milliammeter indicates 6 mA. Calculate the noise factor of the receiver in dB.

Solution
$F = 6$ or $10 \log_{10} 6 = 7.78\ \text{dB}$ (*Ans.*)

(c) Adjacent Channel **Selectivity**

The selectivity of a radio receiver is its ability to select the wanted signal from all the unwanted signals present at the aerial. The selectivity curves given in Fig. 9.5 indicate how well the receiver rejects unwanted signals when the wanted signal is *not* present. This is, of course, not of prime interest since the important factor is the adjacent channel voltage needed to adversely affect reception of the wanted signal. This feature of a receiver is expressed by its adjacent channel response ratio which can be measured using the arrangement shown in Fig. 9.20.

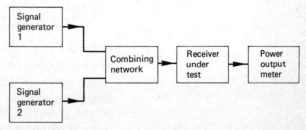

Fig. 9.20 Measurement of radio receiver adjacent channel response ratio

With signal generator 2 producing zero output voltage, signal generator 1 is set to the required test frequency and then is modulated to a depth of 30%. With the input signal voltage at 10 mV, the a.f. gain of the receiver is adjusted to give an audio output power greater than 50 mW but below the

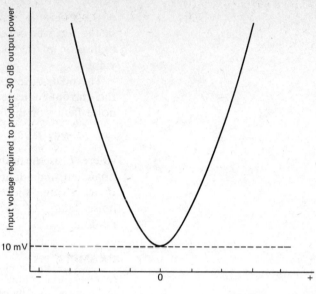

Fig. 9.21 Selectivity characteristic
of a radio receiver

overload point. The modulation of signal generator 1 is then
switched off. Signal generator 2 is then set to a frequency that
is 9 kHz above the test frequency and is modulated to a depth
of 30%. The output voltage of signal generator 2 is then
increased until the audio output power is 30 dB less than the
previous value. The adjacent channel response ratio is the
ratio of these voltages. The measurement can be carried out at
a number of other frequencies and the results plotted (Fig.
9.21).

(d) Image Channel Response Ratio

The image channel response ratio (or rejection ratio) is the
ratio

$$20 \log_{10} \left(\frac{\text{Input voltage at image frequency}}{\text{Input voltage at signal frequency}} \right)$$

to produce the same audio output power. The measurement
can be carried out using the circuit given in Fig. 9.18. The
signal generator and the receiver are each tuned to the test
frequency and the input voltage is adjusted to give an audio
output power of 50 mW. Then, without altering any of the
receiver controls, the frequency of the signal generator is
altered to the image frequency. The input voltage is then
increased until 50 mW audio output power is again registered
by the power meter. The image response ratio is then given by
the ratio of the two necessary input voltages, expressed in dB.

Exercises

9.1. What is meant by the terms *ganging* and *tracking* when applied to the alignment of a superheterodyne receiver? A receiver having an intermediate frequency of 465 kHz is required to tune over a range of 600 kHz to 1800 kHz with a ganged variable capacitor having a range of 320 pF per section. Calculate the values of (*a*) the minimum capacitance needed in the r.f. circuit, (*b*) the inductance required in the r.f. circuit, (*c*) the padding capacitance required in the local oscillator circuit assuming that the minimum value of the capacitance is the same as that found in (*a*), and (*d*) the inductance required to tune the local oscillator. (*C* & *G*)

9.2. With the aid of a block schematic diagram, describe the principle of operation of a superheterodyne receiver suitable for amplitude-modulated sound broadcast reception. What are the reasons for the use of a tuned r.f. amplifying stage? Define the terms (*a*) sensitivity and (*b*) image channel response ratio in relation to the performance of a superheterodyne radio receiver. Why is the image channel response generally lowest when the receiver is tuned to the highest frequency in its range? (*C* & *G*)

9.3. A single superheterodyne receiver with an i.f. of 465 kHz is tuned to an incoming sinusoidal signal of 1000 kHz. Assume the oscillator frequency to be above the signal frequency. What signal appears in the audio output if another unmodulated signal appears in the aerial at a frequency of (*a*) 1932 kHz, (*b*) 1008 kHz, and (*c*) 469 kHz? For each of these frequencies state what factor in the receiver design affects the level of the output signal. Explain the advantages of a double superheterodyne receiver over a single superheterodyne receiver. (*C* & *G*)

9.4. Explain the meaning of the terms and the need for (*a*) ganging, (*b*) padding, (*c*) trimming in a superheterodyne receiver. The capacitor in the r.f. tuned circuit of a superheterodyne receiver varies from 60 pF to 540 pF as the receiver tunes from 1500 kHz to 500 kHz. The receiver has an i.f. of 465 kHz and the local oscillator frequency is above the signal frequency. What value of trimming capacitance should be added to an identical ganged capacitor in the tuning section of the local oscillator to obtain correct tracking at both ends of the frequency range? (*C* & *G*)

9.5. With the aid of a block schematic diagram describe the application of automatic frequency control to an f.m. radio receiver. Why is a.f.c. more necessary at v.h.f. than at lower frequencies? An a.f.c. discriminator produces 1 V of control bias for a frequency error of 50 kHz and the controlled oscillator is shifted by 250 kHz. Calculate the tuning error if the oscillator would have drifted 20 kHz from the correct frequency without a.f.c. (*C* & *G*)

9.6. A superheterodyne receiver has an intermediate frequency of 10.7 MHz, and the local oscillator frequency is above the signal frequency. The receiver covers the band of f.m. carriers spaced at 0.5 MHz intervals between 75 MHz and 97 MHz. Each of these carriers has a modulation index of 5, when the maximum modulating frequency is 15 kHz.

(a) (i) What is the number of f.m. channels in the band? (ii) What is the minimum i.f. bandwidth required? (b) When the receiver is tuned to the carrier at 75 MHz, what is the band covered by (i) the image channel, (ii) the adjacent channel? (c) (i) Which carriers are most susceptible to image channel interference from within the band? (ii) What are the frequencies of the interfering carriers? (*C & G*)

9.7. (i) In what part of a superheterodyne receiver is image rejection provided? (ii) Why is image rejection needed? (iii) Is image rejection better at the high frequency end of the tuning range of the receiver or at the low-frequency end? (iv) Why is an r.f. amplifier not always used? (v) Why is the double-superheterodyne principle often employed? (vi) Quote common intermediate frequencies for such a receiver.

9.8. Draw the block schematic diagram of a superheterodyne receiver. Discuss the functions of each block. What are the advantages of using integrated circuits in a receiver? State some of the functional circuits which are currently available in i.c. form and draw a block diagram for a receiver using some, or all, of these i.c.s.

9.9. A superheterodyne receiver has $f_{if} = 470$ kHz. It is tuned over the frequency band 500–1500 kHz. What range of frequencies must its local oscillator cover? What are (i) the lowest and (ii) the highest image frequencies? How can the image frequencies be suppressed? Draw a typical circuit, including waveband switching.

9.10. A superheterodyne radio receiver is tuned to 1.2 MHz and its local oscillator then operates at 1665 kHz. What is its intermediate frequency? Discuss the reasons leading to the choice of such a frequency for this purpose.

9.11. Draw the block diagram of a superheterodyne receiver. Explain its operation. A receiver tunes to the band 90–100 MHz and has an i.f. of 10.7 MHz. Calculate (i) the range of frequencies over which the local oscillator is tuned and (ii) the maximum and minimum image frequencies. Which of these two frequencies are suppressed most efficiently?

9.12. Explain the meanings of the following terms used with radio receivers: (a) selectivity, (b) sensitivity, (c) image channel response ratio. For each, state which part of the receiver determines the characteristic.

Draw the block diagram of a double superheterodyne receiver; list the function of each block. Quote typical figures for the first and second intermediate frequencies, and hence determine the frequencies of the two local oscillators when a 6 MHz signal is received.

9.13. (a) What is meant by the terms (i) ganging, (ii) tracking when applied to the alignment of a superheterodyne receiver?

(b) A receiver having an intermediate frequency of 465 kHz is required to tune over a range 550 kHz to 1650 kHz with a ganged capacitor which can be varied by 320 pF. Calculate the values of

(i) the minimum capacitance required in the r.f. circuit

(ii) the self-inductance required in the r.f. circuit

(iii) the padding capacitance required in the local oscillator section, assuming that the minimum value of the variable capacitor is that determined in (i)

(iv) the self-inductance required to tune the local oscillator circuit. (*C & G*)

Short Exercises

9.14. In a v.h.f. receiver the local oscillator frequency is 140 MHz and the first intermediate frequency is 10.7 MHz. Calculate (i) the signal frequency (ii) the image channel frequency. Assume the signal frequency is higher than the oscillator frequency.

9.15. Briefly explain the advantages of double superheterodyne operation of a receiver over single-superheterodyne operation.

9.16. State the factors affecting the sensitivity of a radio receiver.

9.17. List the reasons why the local oscillator in a superheterodyne communications radio receiver should be of high frequency stability.

9.18. Why is it usual to provide automatic gain control for an a.m. radio receiver? Why is a.g.c. not always applied to a f.m. receiver?

9.19. What is the function of the limiter in an f.m. receiver? In what stage of the receiver might limiting be provided?

9.20. What is meant by squelch or muting as applied to radio receiver and why is it often applied to communication receivers? Why is it not applied to a.m. broadcast receivers?

9.21. What is meant by the terms ganging and three-point tracking when applied to a superheterodyne radio receiver?

9.22. List the factors which influence the choice of intermediate frequency for a radio receiver.

9.23. Why is it desirable for radio receivers operated at frequencies above about 3 MHz to be provided with r.f. gain?

9.24. An f.m. receiver which is tuned to a frequency of 93 MHz is found to receive interference from a strong signal which has a frequency of 87.65 MHz. What is the name for this type of interference? *(part C & G)*

10 Radio Receiver Circuits

Introduction

The principles of operation of the superheterodyne radio receiver have been discussed in the previous chapter by considering the various sections of the receiver as blocks performing particular functions. The circuits of the various kinds of detector have been discussed in Chapter 2 while radio- and audio-frequency amplifiers and oscillators are covered in a companion volume [EIII]. In this chapter the circuitry and principles of operation of some further radio receiver circuits will be given, namely mixers, crystal and ceramic filters, automatic gain control, and squelch or muting arrangements.

Mixers

A MIXER *is a circuit whose function is to translate a signal from one frequency band to another.* If a mixer incorporates an oscillator it is often known as a *frequency changer.* There are two basic methods by which mixers operate: either the signal and the output of a local oscillator are *added* together and then applied in series with a square-law device, or the two signals are *multiplied* together in a multi-electrode valve or a dual-gate m.o.s.f.e.t.

An additive mixer must include a device having a non-linear input-voltage/output-current characteristic; this device may be a diode or a suitably biased f.e.t. or transistor.

The output current of the non-linear device contains components at the following frequencies:

(a) The signal frequency f_s.
(b) The sum and difference $f_o \pm f_s$ of the signal and the local oscillator frequencies.
(c) The local oscillator frequency f_o.

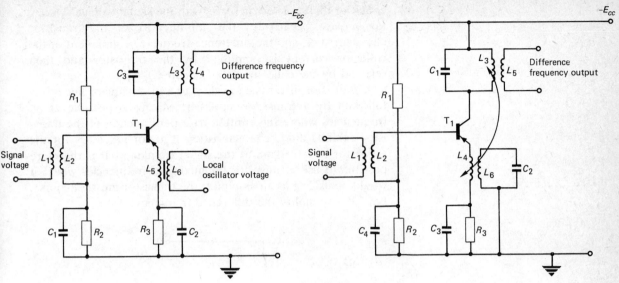

Fig. 10.1 Transistor mixer

Fig. 10.2 Self-oscillating mixer

(*d*) The sum and difference of the local oscillator frequency and the frequencies of all the other components present at the input terminals of the receiver.

(*e*) Various intermodulation and harmonic frequencies.

The output circuit of the mixer is usually tuned to select the difference frequency $f_o - f_s$ and to reject all other components.

The non-linear device may consist of a semiconductor diode or of a suitably biased f.e.t. or transistor. One possible TRANSISTOR MIXER circuit is shown in Fig. 10.1. The transistor is biased with a low collector current so that it is operated on the non-linear part of its characteristics. The local oscillator signal is introduced into the base-emitter circuit via the inductances L_5 and L_6 and the signal voltage is inserted in series with the oscillator voltage via inductors L_1 and L_2. The collector current then contains the wanted difference frequency component plus various other components. The collector circuit is tuned so that the wanted component is selected and all other frequencies are rejected.

Many transistor radio receivers employ a SELF-OSCILLATING MIXER, or frequency changer, of the type shown in Fig. 10.2, because the circuit is cheaper for a comparable performance. Inductors L_3, L_4 and L_6, capacitors C_1 and C_2, and the transistor form the oscillator part of the circuit. Energy is fed from the collector circuit to the L_6-C_2 circuit, which is tuned to resonate at the desired frequency of oscillation. The oscillatory current set up in inductor L_6 induces a voltage, at the oscillation frequency, into the emitter circuit of the transistor in series with the signal voltage. Mixing takes place because of the non-linearity of the transistor characteristics, and the difference-frequency component of the collector current is amplified by the transistor and then selected by the collector tuned circuit C_1-L_3.

A transistor mixer is equivalent to a diode square-law mixer followed by a transistor amplifier. Mixing is possible at all frequencies where the emitter–base p–n junction of the transistor exhibits diode characteristics; it is not necessary for the transistor to have gain at the signal frequency. It is therefore possible to use a transistor for mixing at frequencies where it would be useless as an amplifier; the transistor must, of course, be able to amplify the difference frequency.

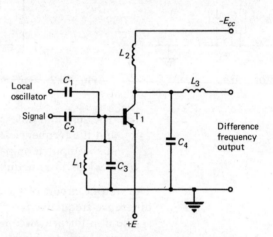

Fig. 10.3 A v.h.f. mixer

The two previous mixer circuits have both introduced the local-oscillator voltage into the emitter circuit. Other methods are possible, and Fig. 10.3 shows the circuit of a V.H.F. MIXER in which both signal and local-oscillator voltages are fed into the base circuit. L_1 and C_3 tune the input circuit of the mixer to the signal frequency, and C_1 and C_2 match the local-oscillator and signal-frequency circuits to the mixer. L_2 is provided to prevent alternating currents from passing into the collector supply, and C_4, L_3 match the mixer to the output circuit and select the difference-frequency component of the collector current.

Important features of a mixer are (*a*) its conversion conductance, given by

$$\left(\frac{\text{difference frequency component of the output current}}{\text{r.f. input signal voltage}}\right)$$

and (*b*) its cross-modulation performance. Cross-modulation is the transfer of the modulation of an unwanted carrier onto the wanted carrier, and it can occur in a mixer if its mutual characteristic includes a cubic term. The mutual characteristic of a field-effect transistor more nearly approaches the ideal

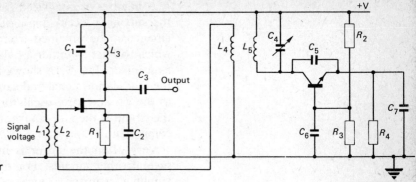

Fig. 10.4 Square-law f.e.t. mixer

square law and consequently f.e.t.s are increasingly employed
as mixers in modern circuitry. A typical f.e.t. mixer circuit is
shown in Fig. 10.4 together with an example of an oscillator
circuit which is often used at v.h.f.

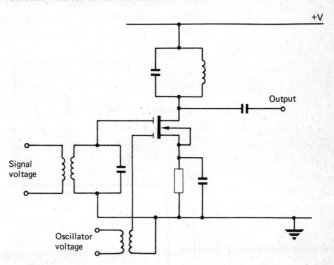

Fig. 10.5 Dual-gate m.o.s.f.e.t. mixer

MUTIPLICATIVE MIXING can be achieved by applying
the signal and the local oscillator voltages to the two inputs of
a dual-gate m.o.s.f.e.t. and Fig. 10.5 shows a dual-gate
m.o.s.f.e.t. mixer circuit. The oscillator voltage is applied to
gate 1 of the m.o.s.f.e.t. and the signal voltage is applied to
gate 2. The oscillator voltage varies the mutual conductance of
the m.o.s.f.e.t. and, since $I_d = g_m V_s$, the alternating drain current is proportional to the product of the instantaneous values
of the signal and the local oscillator voltages. The drain
current contains components at a number of different frequencies amongst which is the wanted difference $f_o - f_s$ component.
Mutiplicative mixing possesses two advantages over additive
mixing; firstly, the signal and local oscillator circuits are isolated from one another which prevents *oscillator pulling* taking
place and, secondly, its conversion conductance is higher. On
the other hand multiplicative mixing is a noisier process.

A number of integrated circuits are available which include the mixing process amongst their functions. An integrated circuit can be obtained from several different manufacturers which will act as either a balanced modulator/demodulator or as a mixer. Fig. 3.11 shows a possible circuit arrangement if the modulating signal and carrier inputs are read, respectively, as the r.f. and local oscillator voltages. The output voltage of the circuit is then, of course, the required difference frequency component.

Very often the mixer is included in the same package as several other circuits. For example, one i.c. contains an r.f. amplifier, a mixer, a local oscillator, an i.f. amplifier and an a.g.c. detector. With any linear integrated circuits any necessary inductors and capacitors and large-value resistors must be provided externally. Fig. 10.6 shows a simplified example of this. The gain/frequency characteristics of the r.f. and i.f. amplifiers and the mixer are determined by the tuned circuits shown. The a.g.c. voltage line must have a particular time constant and this is provided by capacitor C_7.

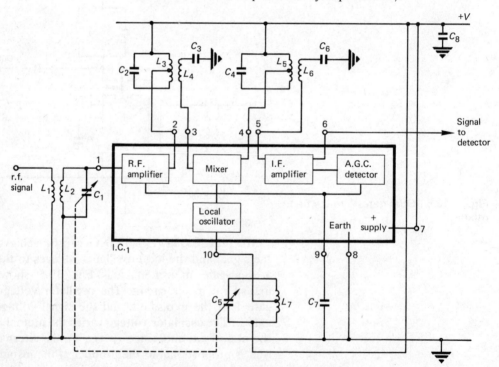

Fig. 10.6 Use of an integrated circuit in a radio receiver

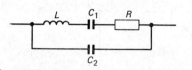

Fig. 10.7 Electrical equivalent circuit of a piezo-electric crystal

Crystal and Ceramic Filters

The principles of operation of piezo-electric cystals have been discussed in another volume [TSII] and in this section of the book some practical radio applications will be considered.

The electrical equivalent circuit of a piezo-electric crystal is shown in Fig. 10.7; the inductance L represents the inertia of the crystal, capacitance C_1 represents the crystal's compliance (1/stiffness) and the resistance R provides losses which are equivalent to the frictional losses of the crystal. Lastly, the shunt capacitor C_2 is the actual electrical capacitance of the crystal. A series-parallel circuit of this kind has two resonant frequencies, one is the resonant frequency of the series circuit L-C_1-R and the other is the frequency at which parallel resonance occurs between shunt capacitor C_2 and the net (inductive) reactance of the series arm. Obviously, the parallel-resonant frequency is higher than the frequency of series resonance. The crystal will pass, with little attenuation, all frequencies in between the series and parallel resonant frequencies. Often this bandwidth is too narrow and when this is the case it can be widened by connecting an inductor of suitable value in series with the crystal. The added inductance has this effect because it will reduce the series-resonant frequency of the crystal without affecting its parallel-resonant frequency. Usually, this technique is only employed at the lower frequencies since when the centre frequency is high the bandwidth, as a percentage of the centre frequency, is likely to be wide enough.

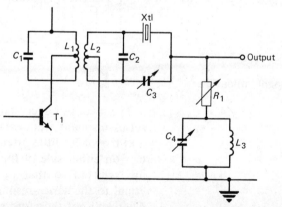

Fig. 10.8 Crystal filter

A CRYSTAL FILTER circuit is shown in Fig. 10.8. The crystal is chosen to be one whose series-resonant frequency is equal to the required passband's lowest frequency. The parallel-resonant frequency of the crystal, and hence the upper passband frequency, is adjusted to the required figure by means of the variable capacitor C_3. The selectivity of the circuit is partly determined by the load impedance of the filter and this can, to some extent, be varied by adjustment of C_4 and/or R_1.

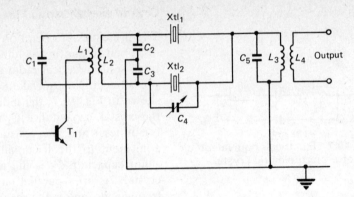

Fig. 10.9 A two-crystal filter

A wider bandwidth filter can be obtained by the use of two crystals connected as shown in Fig. 10.9. The series-resonant frequencies of the crystals are chosen to differ from one another by a frequency equal to the wanted passband; this may be only a few hundred hertz if telegraphy signals are to be received or about 3 kHz for the reception of an s.s.b. signal. Capacitor C_4 is provided to allow for fine adjustment of the bandwidth provided. Even better selectivity characteristics can be obtained, but at greater expense, if four crystals are connected to form a lattice network (see Fig. 10.10).

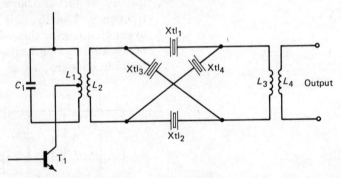

Fig. 10.10 Crystal lattice filter

The two series crystals are chosen so that their common series-resonant frequencies lie within the required passband and they offer little attenuation to a narrow band of frequencies on either side of this frequency. The two parallel crystals are selected so that their parallel-resonant frequencies are equal to the series-resonant frequencies of the series crystals. They will not therefore shunt signals in the wanted passband. At frequencies outside the required passband the series crystals will have a high impedance and the parallel crystals will have a low impedance, and the network will offer considerable attenuation.

Complete crystal filter modules, suitable for use in the i.f. amplifier stage of a radio receiver, can be purchased from various manufacturers.

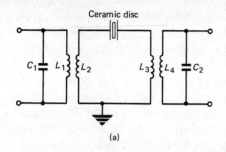

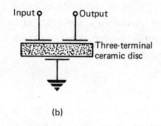

Fig. 10.11 Ceramic filter

The piezo-electric effect is also obtained when a ceramic disc has electrode plates mounted on each of its two faces, the resonant frequencies and selectivity characteristic being determined mainly by the shapes and dimensions of the electrodes and the disc. The make-up of a CERAMIC FILTER is shown in Fig. 10.11a; this type of filter, available as a complete sealed unit, is usually manufactured for use at one of the standard a.m. broadcast receiver intermediate frequencies. The input and output tuned circuits are both arranged to be resonant at the centre frequency of the desired passband. Ceramic filters for use at the higher frequencies, such as the standard 10.7 MHz intermediate frequency of many v.h.f. receivers, are usually of the three-electrode type shown in Fig. 10.11 b.

Both crystal and ceramic filters possess the following advantages over conventional tuned circuits: compact size, high stability and insensitivity to external magnetic fields. The ceramic filter is cheaper than the crystal filter and requires no tuning or alignment but, on the other hand, its passband loss is greater, it has a worse temperature stability, and its selectivity is not as great. Usually, ceramic filters are found in domestic broadcast radio receivers, particularly in conjunction with integrated circuits, and crystal filters are used in communication receivers, for example as a channel filter in an i.s.b. receiver. A simplified example of the use of a ceramic filter is given in Fig. 10.12; $I.C._1$ is an r.f. wideband amplifier whose selectivity characteristic is shaped to the required i.f. amplifier response by the input and output ceramic filters F_1 and F_2. All the necessary decoupling capacitors are supplied by external components because of the relatively high (0.01 μF) values required.

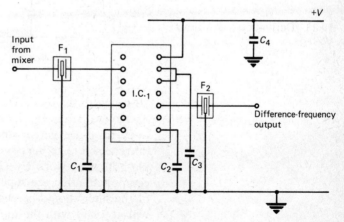

Fig. 10.12 Wideband amplifier using ceramic filters

Automatic Gain Control

The signals arriving at the input terminals of a radio receiver are subject to continual fading, and unless *automatic gain control* (a.g.c.) is used, the volume control will require continual adjustment to keep the output of the receiver more or less constant. *The function of an a.g.c. system is to vary the gain of a receiver to maintain a reasonably constant output power even though there are large variations in the input signal level.* Thus the gain of the receiver must be reduced by the a.g.c. system when a large-amplitude input signal is received, and increased for a small input signal. The variation in the receiver gain also serves to prevent the output level changing overmuch as the receiver is tuned from one station to another, and it also avoids a.f. amplifier distortion caused by overloading on larger input signals.

Simple A.G.C.

The voltage appearing across the load resistor of a diode detector contains a direct component, the magnitude of which is directly proportional to the amplitude of the carrier voltage. This direct voltage is available for use as the a.g.c. voltage and can be fed to the controlled stages in the manner shown in Fig.

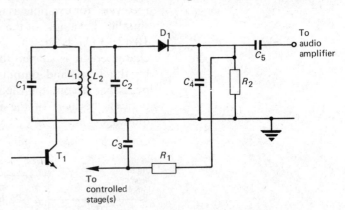

Fig. 10.13 Derivation of simple a.g.c. voltage

10.13. With the diode D_1 connected as shown, the d.c. voltage developed across the load resistor R_2 is positive with respect to earth; if a negative voltage is required the diode D_1 should be reversed. The a.g.c. voltage is fed to the controlled stages via a filter network $C_3 - R_1$ to remove the various alternating components that are superimposed upon it. The time constant of the filter should be chosen to ensure that the a.g.c. voltage will not vary with the modulation evelope but will respond to the most rapid fades to be expected. Typically, time constants of 0.05 to 0.5 seconds are used.

Delayed A.G.C.

A delayed a.g.c. system will not produce any a.g.c. voltage until the carrier level at the output of the detector is greater than some predetermined value. This means that the diode that produces the a.g.c. voltage must be biased into its non-conducting state by a bias, or *delay*, voltage of suitable magnitude. It is obvious that the signal diode cannot be biased into non-conduction and so it is necessary to use a separate diode for the a.g.c. voltage.

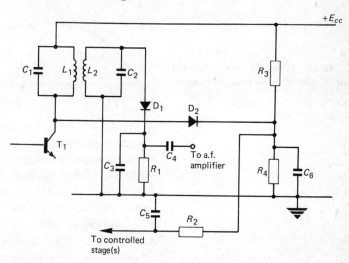

Fig. 10.14 Derivation of delayed a.g.c. voltage

A large number of different circuits have been produced for delayed a.g.c. systems and Fig. 10.14 shows a typical circuit. The signal diode D_1 is operated as a normal signal diode with a load resistor R_1. The a.g.c. diode D_2 has a positive bias voltage $VR_4/(R_3 + R_4)$ at its cathode and will not conduct until the signal voltage appearing at its anode is greater than the bias voltage. When diode D_2 conducts, the a.g.c. voltage is developed across R_4 and is fed to the controlled stages via the filter network $C_5 - R_2$. The a.g.c. diode is supplied from the collector of T_1 to obtain as large a voltage as possible.

F.M. Receiver Main A.G.C.

A main a.g.c. loop may sometimes be applied to an f.m. receiver to ensure that the signal level to the input of the limiter stage is always large enough for the limiting action to take place. The a.g.c. voltage can be obtained from the d.c. load capacitor of a ratio detector or, if a ratio detector is not used, from the limiter stage itself. Alternatively, a separate a.g.c. diode can be employed.

Applying the A.G.C. Voltage to the Controlled Stage

The current gain of a bipolar transistor is a function of its emitter current and Fig. 10.15 shows a typical gain/emitter-current characteristic. At low values of emitter current the current gain of a transistor increases with increase in its emitter current in a more or less linear manner. The converse is true at high values of emitter current; an increase in the emitter current produces a fall in the current gain. Two

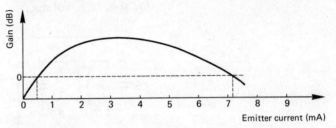

Fig. 10.15 Gain/emitter current characteristic of a bipolar transistor

methods are therefore available for varying the gain of an a.g.c. controlled stage in a radio receiver: increasing the gain by *increasing* the emitter current is known as *reverse a.g.c.*, while increasing the gain by decreasing the emitter current is known as *forward a.g.c.* In either case the emitter current is most easily controlled by variation of the base-emitter forward bias voltage of the transistor since minimum power is then taken from the a.g.c. line.

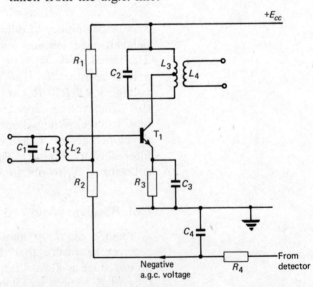

Fig. 10.16 Reverse a.g.c.

Fig. 10.16 shows the application of REVERSE A.G.C. to a transistor tuned amplifier. The negative polarity a.g.c. voltage determines the forward bias voltage applied to the transistor T_1. An increase in the carrier level at the input to the detector stage will make the a.g.c. voltage become more negative. The base potential of T_1, relative to earth, will become less positive and the transistor will conduct a smaller emitter current. The voltage gain of the amplifier will therefore fall and the carrier level at the detector will be reduced, tending to compensate for the original increase.

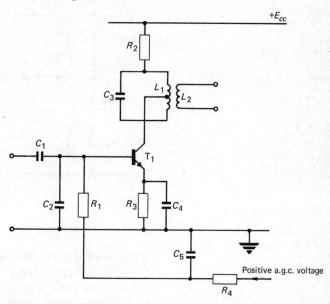

Fig. 10.17 Forward a.g.c.

When FORWARD A.G.C. is to be applied to an amplifier, a resistor is connected in series with the collector tuned circuit to increase the gain variation produced by a given a.g.c. voltage. A typical forward a.g.c. circuit is given in Fig. 10.17. When the positive a.g.c. voltage increases, because of an increase in the received carrier voltage, the forward bias of the transistor is also increased. The transistor conducts a larger current and so its current gain falls; the fall in gain is accentuated by the collector-emitter voltage also falling because of the increased voltage drop across the series resistor R_2.

The relative merits of reverse and forward a.g.c. are as follows. Reverse a.g.c. controls the gain of a stage by varying its emitter current; to reduce the gain the emitter current must be reduced and as a result the output resistance of the transistor increases. This, in turn, reduces the damping effect of the transistor on the collector tuned circuit and so reduces the bandwidth of the stage. The reduction in emitter current, and hence in the collector current, also has the effect of reducing the signal-handling capability of the stage—at the very time it is being called upon to handle a signal of larger amplitude. Conversely, with forward a.g.c. a decrease in the voltage gain of a stage is obtained by increasing the emitter and collector currents and is therefore associated with an increase in both the bandwidth and the signal handling capacity of the stage. Also, the d.c. collector current taken from the supply is greater

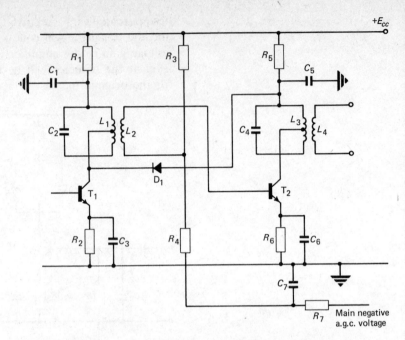

Fig. 10.18 Auxiliary a.g.c.

with forward a.g.c. than with reverse a.g.c. and this may prove
to be an embarrassment with battery-operated equipment. Fig.
10.18 shows one way in which auxiliary a.g.c. can be applied
to a receiver. The circuit operation is left as an exercise for the
reader.

In an s.s.b./i.s.b. communication receiver, the a.g.c. system is
often operated from a low-level pilot carrier which is filtered
off from the wanted signal. If there is no pilot carrier, it is
possible to derive an a.g.c. voltage from the received signal
itself, either at the i.f. amplifier or the a.f. amplifier stages.

Automatic Frequency Control

The direct voltage needed to activate the automatic frequency control (a.f.c.) system of a frequency-modulation radio
receiver can be derived from the audio load capacitor of the
ratio detector. Fig. 10.19 shows the essentials of a possible
circuit arrangement. When the mean value of the intermediate
frequency of the receiver is correct, the voltage appearing
across the audio load capacitor C_4 has zero d.c. component
and the varactor diode has the capacitance determined by the
applied bias voltage. If the average value of the intermediate
frequency should drift from its nominal value, the voltage
developed across the audio load capacitor will have a direct
component. This direct voltage is applied to the varactor diode
to augment or oppose, depending on its polarity, the bias
voltage and so vary the diode capacitance. This variation in the

diode capacitance alters the frequency of the local oscillator in the direction necessary to reduce the error in the intermediate frequency. Capacitor C_3 and resistor R_2 act as a low-pass filter to remove all alternating voltages from the a.f.c. line.

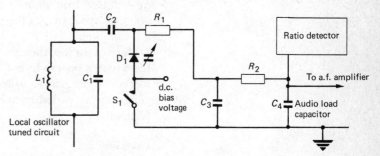

Fig. 10.19 Application of a.f.c. voltage in a radio receiver

Squelch or Muting

A sensitive radio receiver will have a very high gain between its aerial terminals and its detector stage. When it is not receiving a carrier, and so develops zero a.g.c. voltage, its full voltage gain will be made available to amplify the noise unavoidably present at its input stage. As a result there will be a high noise level at the output of the audio amplifier which may cause considerable annoyance to the person operating the receiver. To reduce or eliminate this annoyance, the audio-frequency amplifier can be cut off, or its voltage gain severely reduced, whenever there is no input carrier signal; this is the function of a squelch, or a muting, circuit.

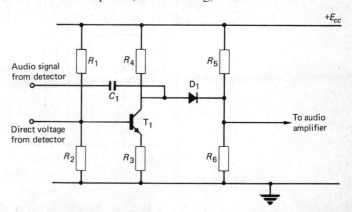

Fig. 10.20 Squelch circuit

A variety of different circuits are capable of providing the squelch action and an example of a squelch circuit is given in Fig. 10.20. When a carrier voltage is present at the detector input, a direct voltage, proportional to the carrier level, is applied to the base of transistor T_1. The polarity of this direct voltage is such that T_1 is turned off and the collector potential of the transistor rises to $+E_{cc}$ volts. The diode D_1 conducts and the audio-frequency output voltage of the signal detector is able to pass to the a.f. amplifier. With zero carrier voltage at the detector input, transistor T_1 is able to conduct and its collector potential falls to a lower positive value than is present at the junction of resistors R_5 and R_6. Diode D_1 is now biased into its non-conducting state and prevents noise voltages appearing at the detector output and passing onto the a.f. amplifier.

Very often it is thought desirable for operational reasons for the squelch system to not cut off the a.f. amplifier but, instead, to reduce its gain to a low value. The output noise level can generally be varied by means of an adjustable squelch circuit control.

Exercises

10.1. (*a*) From what stage in a superheterodyne receiver is a.g.c. derived? (*b*) To what stages in a superheterodyne receiver can a.g.c. be applied? (*c*) Why is delayed a.g.c. preferred to simple a.g.c.? (*d*) List the components in the circuit shown in Fig. 10.21 which contribute to the a.g.c. action and indicate the function of each. How does this circuit compensate for a change in input signal level? (*C & G*)

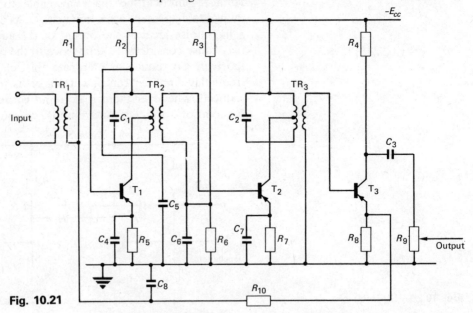

Fig. 10.21

10.2. (*a*) Sketch the circuit of a transistor frequency changer stage with a following i.f. amplifier stage suitable for a medium-wave superheterodyne receiver. Describe the operation of these stages and show which elements in the circuit determine each of the three frequencies involved. (*b*) What is a typical gain for a single stage of the i.f. amplifier in a medium-wave amplitude-modulation broadcast receiver? (*C & G*)

10.3. (*a*) Give reasons for the use of a.g.c. in a superheterodyne receiver designed for the reception of (i) a.m. signals (ii) f.m. signal. (*b*) Draw and explain a circuit that shows how the a.g.c. voltage can be applied to a transistor i.f. amplifier.

10.4. With the aid of a circuit diagram explain the operation of the mixer and local oscillator of a transistor superheterodyne a.m. sound receiver. If the input to a mixer consists of two sinusoidal waves, at what frequencies do the significant components of the output occur? If a wanted sinusoidal wave and a sinusoidal wave at the image frequency are mixed with the local oscillator frequency of a receiver, explain why a large number of the output components of the mixer do not appear at the audio output. What does appear at the audio output? What appears at the output if the unwanted sinusoid is shifted by 2 kHz.?

(C & G)

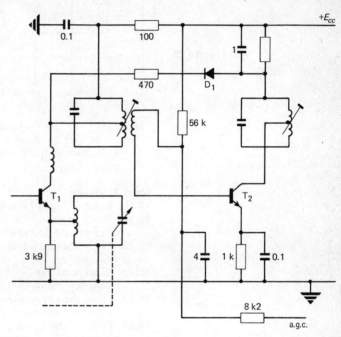

Fig. 10.22

10.5. With reference to Fig. 10.22: (*a*) Why are D_1 and its associated components included in the circuit? (*b*) Explain fully the operation of the section of the circuit which includes D_1.

(C & G)

10.6. With the aid of a graph explain the meaning and purpose of delayed a.g.c. Draw the circuit diagram of the i.f. amplifier and detector stages of a transistor superheterodyne receiver to show how simple a.g.c. is derived and applied. Why must the time constant in the a.g.c. feedback path be about 0.1 seconds?

(C & G)

10.7. (*a*) Draw the circuit diagram of a ratio detector which will produce a suitable output to control the frequency of the receiver oscillator section. (*b*) Explain how the a.f.c. voltage is produced by the circuit given in (*a*) and describe how this voltage maintains the local oscillator at the required frequency.

10.8. With the aid of a simplified circuit diagram explain how a.f.c. may be included in the design of a transistorized f.m. receiver. Include in the diagram a switch which will enable the a.f.c. to be switched out of circuit for tuning purposes and explain why this is desirable.

(C & G)

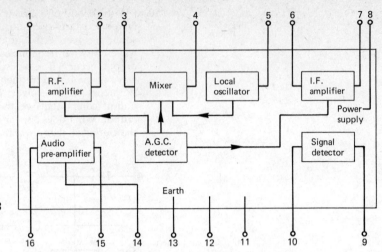

Fig. 10.23

10.9. Fig. 10.23 shows the block schematic diagram of an integrated circuit that has been designed for use in a radio receiver. Draw a diagram showing the **external** components which are necessary.

Short Exercises

10.10. Redraw Fig. 10.6 using ceramic filters.

10.11. What is meant by the term squelch as applied to a radio receiver and describe how it works.

10.12. Draw the circuit of a crystal i.f. filter and describe how it works.

10.13. What is meant by automatic frequency control? Why can the frequency error never be completely eliminated by a.f.c.?

10.14. Draw the circuit diagram of a dual-gate m.o.s.f.e.t. mixer and explain its operation.

10.15. Draw a circuit diagram showing how reverse a.g.c. may be applied to a p-n-p transistor stage.

10.16. Draw a circuit diagram of a mixer that uses an integrated circuit. List the advantages of using integrated circuits.

11 Wideband Line and Radio Systems

Introduction

The public transmission network of a country is used for the communication of many kinds of information such as commerical quality speech, telegraphy, data signals and sound/television signals for the broadcasting authorities. The network will contain audio-frequency circuits, pulse-code modulation systems and multi-channel systems routed over both coaxial cable and microwave *radio-relay systems*. The choice of transmission system for a particular route is determined by a careful consideration of all the relevant factors such as the required transmission performance, the economics involved, and the nature of the terrain to be covered. Local lines and junctions whose length is less than about 16 km mainly operate as audio-frequency circuits, sometimes with two-wire amplification, although increasingly junctions use pulse code modulation. Longer-distance trunks are nearly all routed over one or more multi-channel telephony systems.

International links between nearby countries are also established using both land coaxial and radio relay systems. Links between the United Kingdom and the Continent are mainly routed over submarine coaxial cables and, in the case of England–France, over radio relay systems also. Longer-distance intercontinental telephone circuits are routed over communication satellite systems and submarine coaxial systems, augmented when and where necessary by high-frequency radio links using the sky wave mode of propagation.

One other form of fixed radiocommunication system is used on routes where the terrain is too hostile, geographically or politically, for a coaxial cable (land or submarine) or a radio relay system to be used and the use of a communication satellite cannot be economically justified. Such *tropospheric scatter* systems can provide about a hundred speech circuits

207

over a distance of up to about 600 km. The one example of the use of such a system in the United Kingdom is the links between oil rigs in the North Sea and the mainland of Scotland.

Land Coaxial Systems

Coaxial multi-channel telephony systems play an important part in both national and international telecommunication networks. A number of different types of coaxial system are in current use in the United Kingdom and these are listed in Table 11.1.

Table 11.1

Number of channels	Bandwidth in MHz
900	4
2 700	12
10 800	60

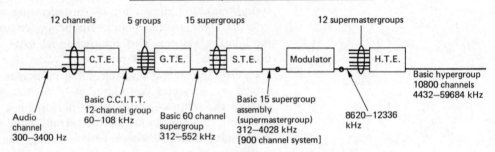

Fig. 11.1   Assembly of a hypergroup
C.T.E. channel translating equipment
G.T.E. group translating equipment
S.T.E. supergroup translating equipment
H.T.E. hypergroup translating equipment

A coaxial system is built up by assembling the appropriate number of 12-channel C.C.I.T.T. groups [TSII]. Fig. 11.1 shows how the 900 and 10 800 channel coaxial systems are assembled. Five 12-channel groups are combined by the group translating equipment to form a 60-channel supergroup occupying the frequency band 312–552 kHz. In the next stage of assembly 15 supergroups are combined to form a 900-channel system known as a supermastergroup. The 900-channel system is sometimes transmitted to line in its own right. To obtain a larger capacity 12 MHz or 60 MHz coaxial system, a number of supermastergroups are used to modulate different carrier frequencies in order to assemble them alongside one another in the appropriate frequency band. In the example shown in the figure, 12 supermastergroups are combined by a hypergroup translating equipment to form a 10 800-channel hypergroup. The carrier frequencies used at each stage in the assembly of the system are given in the preceding volume [TSII].

Single-sideband amplitude modulation is used at each stage of frequency translation at both the transmitting and receiving ends of the system. The building block approach to the assembly of a system is used since it reduces the number of modulators and different filters which are needed. At each stage of frequency translation at the receiving end of the system, a locally-generated carrier must be fed into the balanced modulator. The C.C.I.T.T. recommendation is for the re-inserted carrier to be accurate to within ±2Hz of the carrier frequency suppressed at the other end of the system. The highest carrier frequency used in the 60 MHz system is 68 200 kHz and so the necessary frequency stability is ±3 parts in 10^8. To achieve such a high stability it is necessary to synchronize the carrier frequency generating equipment in a repeater station to a *pilot carrier* which is transmitted with the multiplex signal. Modern carrier generating equipments are extremely accurate and stable, and the generating frequencies need only be compared with the pilot carrier at regular intervals of time. The 60 MHz multiplex signal is amplified and equalized at 1.5 km intervals along the length of the line. At the receive end of the system, the signal receives further amplification and equalization before it is applied to the translating equipment to be broken down into its individual channels.

Radio Relay Systems

Radio relay systems using line-of-eight transmissions in the u.h.f. and s.h.f. bands can provide a large number of telephone channels and/or a television signal. Table 11.2 lists the systems that are currently operated in the United Kingdom.

Table 11.2

Frequency band MHz	Number of r.f. channels	Number of telephone channels	Television
1700–1900	1	960	yes
1900–2300	4	960	no
3790–4200	4	1800	no
5925–6425	6	1800	no
6425–7110	12	960	yes
10 700–11 700	10	960	yes

Radio relay systems are operated in the upper part of the u.h.f. band and in the s.h.f. band because it is then possible to provide a bandwidth of several megahertz. A wideband system is needed to accommodate several hundreds of telephone channels or a television channel. At these frequencies high gain aerials of reasonable physical size are available which make it possible to use transmitted powers of only a few watts. Further, the high directivity obtained minimizes interference from and to other systems.

The BASEBAND SIGNAL (the multiplex signal produced by a coaxial multi-channel system or a television camera) is used to frequency-modulate a 70 MHz carrier wave. The modulated signal is then translated to the allocated frequency band. Frequency modulation is used in preference to amplitude modulation for two main reasons. Firstly, an f.d.m. telephony baseband signal requires a very linear transfer characteristic (output/input) for all the sections of a relay system if inter-channel crosstalk, arising from intermodulation, is to be avoided. The necessary linearity is much easier to provide if frequency modulation is used. Secondly, the use of frequency modulation provides an increase in the output signal-to-noise ratio of the system. The C.C.I.R. recommend the frequency deviations to be used; these are 140 kHz per channel for both 960 and 1800 channels systems.

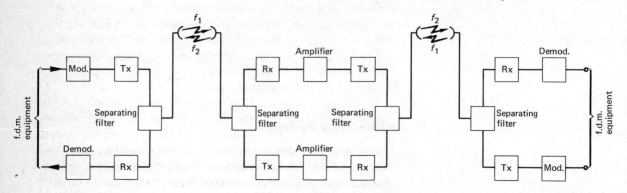

Fig. 11.2 Radio relay system (Rx receiver, Tx transmitter)

The basic block diagram of a RADIO RELAY SYSTEM is given in Fig. 11.2. Only one relay station has been shown but usually several will be used, the actual number depending upon the length of the route.

At the transmit terminal, the baseband signal is pre-emphasized and is then used to frequency-modulate a 70 MHz carrier. The modulated wave is then shifted to the allocated part of the frequency spectrum and amplified before it is radiated by the parabolic dish aerial. At the relay station, the received signal has its frequency changed back to 70 MHz before it is amplified and then shifted back to the frequency band to be used over the next section of the route. At the receiving end of the system, the signal is reduced in frequency to 70 MHz before it is demodulated and, finally, the baseband signal is passed through the de-emphasis circuit to be restored to its original amplitude relationships.

Fig. 11.3 shows in more detail the equipment used at the TRANSMITTING end of the system. The input baseband signal can be a 960 or a 1800 channel f.d.m. telephony system or a television signal. The telephony signal occupies a bandwidth of 60 kHz–4028 kHz or 316–8248 kHz, i.e. 3.97 MHz or 7.93 MHz, while the television signal bandwidth is 8 MHz.

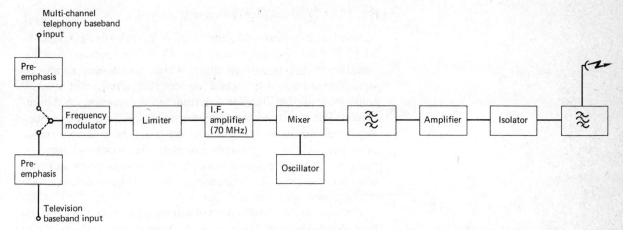

Fig. 11.3 Transmitter in a radio relay system

The baseband signal is pre-emphasized before it frequency-modulates a 70 MHz carrier wave. Different de-emphasis networks are used for telephony and for television signals (characteristics shown in Fig. 2.10) but otherwise the same items of equipment are used. The frequency-modulated signal is amplitude limited to remove any amplitude modulation present and is then amplified before it is translated to the required frequency band by the mixer stage and its following low-pass filter. This filter selects one sideband of the mixing process. The translated signal is then further amplified to the required transmitted power level, typically 10 W, and is then passed on to the aerial via an *isolator* and another filter. The isolator is a ferrite device which will only allow signals to pass in one direction and it is used to prevent any unwanted signals picked up by the aerial passing into the transmitting equipment. Any reflected signals caused by mismatch at the aerial terminals will also be stopped from reaching the equipment. The final band-pass filter is provided to band-limit the transmitted signal in order to avoid interference with adjacent systems.

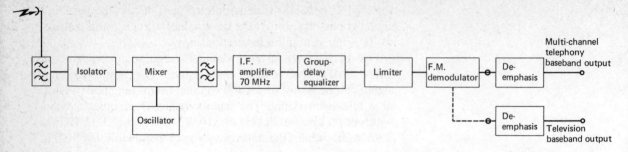

Fig. 11.4 Receiver in a radio relay system

The block schematic diagram of a RADIO RELAY SYSTEM RECEIVER is given by Fig. 11.4. The received signal is selected by the band-pass filter, which reiects any unwanted signals that are also picked up by the aerial, and then is translated to the 70 MHz intermediate frequency of the receiver. The intermediate-frequency signal is amplified, group-delay equalized, and amplitude limited before it is demodulated to obtain the baseband signal. The baseband signal is passed through the appropriate de-emphasis network in order to restore its frequency components to their original amplitude relationships with one another.

The equipment used in a relay station consists of the back-to-back connection, at the intermediate frequency, of a receiver and a transmitter equipment. The output of the limiter in the receiver diagram is at 70 MHz and this is directly connected to the input of the i.f. amplifier shown in the transmitter diagram.

The parabolic reflector aerials used with radio relay systems have the capability to transmit or receive more than one r.f. channel at the same time. It is therefore usual for more than one r.f. channel to be multiplexed onto a single aerial. To improve the discrimination between channels, adjacent (in frequency) channels use alternate planes of polarization. For example, if channel 1 is horizontally polarized, channel 2 will be vertically polarized, channel 3 will be horizontally polarized, and so on. The block diagram of the equipment involved is shown in Fig. 11.5. As before only one relay station is shown but usually there will be several more. A *circulator* is a ferrite device with four input/output terminals; the operation of the device is such that a signal entering one pair of terminals will be directed only to one other pair of terminals—none of the input energy will appear at the other two pairs of terminals.

Both radio relay and coaxial systems are widely used as integral parts of the national telephone network. The two systems have a number of advantages and disadvantages relative to one another which often means that one or the other is best suited for providing communication over a given route.

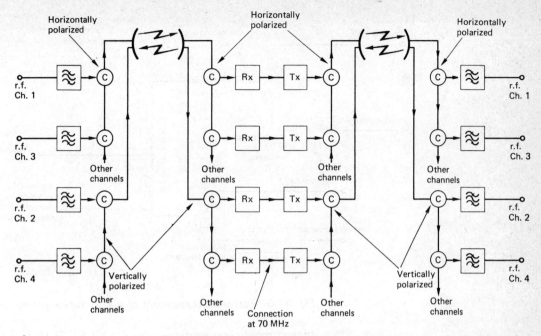

Fig. 11.5 Method of transmitting several r.f. channels over a single radio relay system (C = circulator)

The relative merits of the two systems are listed below:

(a) A radio relay system is generally quicker and easier to provide.
(b) The problems posed by difficult terrain are easier to overcome using a radio relay system.
(c) It is easier to extend the channel capacity of a radio relay system.
(d) Difficulties may be experienced in obtaining suitable (line-of-sight distance) sites for a radio relay system.
(e) When relay station sites have been chosen it may be difficult to gain access to them, whereas coaxial systems usually follow roads and so access is relatively easy.
(f) The transmission performance of a radio relay system is adversely affected by bad weather conditions.

Submarine Cable Systems

Submarine coaxial cable [TSII] is designed for the transmission of wideband signals beneath the seas and oceans. It can provide large numbers of good quality, highly reliable, telephone channels over long distances, for example between Europe and North America.

A single coaxial pair is used for both directions of transmission, transmission in one direction being achieved in one frequency band with transmission in the opposite direction in a

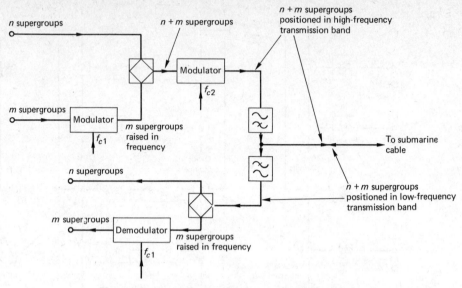

Fig. 11.6 Terminal equipment of a submarine cable system

higher band. For example, the CANTAT II system between the United Kingdom and Canada uses the frequency band 312–6012 kHz in one direction and the frequency band 8000–13 700 kHz in the other. Further, 3 kHz bandwidth telephone channels are provided, instead of the 4 kHz channels customary in land systems, to economize in the use of the frequency spectrum. This means that the bandwidth occupied by a normal supergroup, i.e. 60×4 kHz $= 240$ kHz, can now accommodate 80 telephone channels. Fig. 11.6 shows, in block diagrammatic form, the equipment used at a terminal which is transmitting signals in the higher frequency band and receiving signals in the lower frequency band. A number n of supergroups are assembled in the usual manner (see Fig. 11.1), and are combined in a hybrid coil with a number m of other supergroups that have been translated to a higher part of the frequency spectrum by a stage of modulation. The combined $(m+n)$ supergroups are next positioned in the high-frequency transmission band by another stage of modulation and then applied to the input terminals of the submarine cable. Incoming supergroups are positioned in the lower frequency band and do not require demodulation before being separated into two groups of n and m supergroups.

High-frequency Radio Systems

Submarine cable systems are extremely expensive and their traffic capacity is not sufficient to satisfy the ever-increasing demand for international communication facilities. Some long-

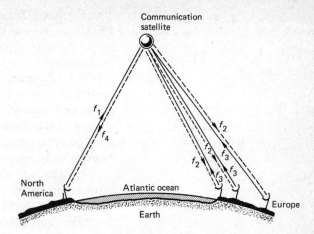

Fig. 11.7 Communication satellite system

distance international circuits are provided in the 3–30 MHz high-frequency band. Modern high-frequency systems can provide a fairly reliable service and for most of the time although sometimes propagation conditions are such that communication by this means is not possible. High-frequency radio systems have the advantages of (a) a relatively low capital cost and (b) flexibility; and it is expected that they will find continued application as a supplement to submarine cable and communication satellite systems.

Communication Satellite Systems

Most of the long-distance international telephone traffic which is not carried by submarine cable systems is routed via a broadband communication satellite system, the basic principle of which is illustrated by Fig. 11.7. The ground stations are fully integrated with their national telephone networks and, in addition, the European ground stations are fully interconnected with one another. Four frequencies are used; the North American ground station transmits on frequency f_1 and receives a frequency f_4, the European stations transmit frequency f_3 and receive frequency f_2. Essentially, the purpose of the communication satellite is to receive the signals transmitted to it, frequency translate them to a different frequency band (f_1 to f_2 or f_3 to f_4), amplify the signals, and then re-transmit them to the ground station at the other end of the link.

Communication satellites which form an integral part of the public international telephone network are operated on a global basis by COMSAT (Communication Satellite Corporation) on behalf of an international body known as INTELSAT

(International Telecommunication Satellite Consortium). The COMSAT system employs communication satellites travelling in the circular equatorial orbit at a height of 35 880 km. This particular orbit is known as the synchronous orbit because a satellite travelling in it appears to be stationary above a particular part of the Earth's surface. Seven satellites are used, positioned around the Earth so that nearly all parts of the surface of the Earth are "visible" from at least one satellite. A large number of ground stations are in use and now number more than a hundred in nearly a hundred different countries.

Each ground station transmits its telephone traffic to a satellite on the particular carrier frequency allocated to it in the frequency band 5.925–6.425 GHz. This is a bandwidth of 500 MHz and allows the simultaneous use of a satellite by more than one ground station. Different ground stations are allocated different carrier frequencies within this 500 MHz band, either permanently or for particular periods of time—depending on the traffic originated by that station. This method of utilizing the capacity of a communication satellite is known as *frequency division multiple access* (f.d.m.a.). Each of the allocated carrier frequencies has a sufficiently wide bandwidth to allow a multi-channel telephony system to be transmitted and, in some cases, a television channel. The number of telephony channels thus provided varies from 24 in a 2.5 MHz bandwidth to 1872 in a bandwidth of 36 MHz. All the signals transmitted by a satellite are transmitted towards every ground station and each station selects the particular carrier frequencies allocated to it in the band 3.700–4.200 GHz.

Noise and Interference in Communication Systems

The output of any communication system, line or radio, will always contain some unwanted components superimposed upon the desired signal waveform. The unwanted voltages are the result of *noise* and *interference* picked up by or generated within the system. The sources of noise and interference in communication systems are many and have been discussed elsewhere [EIII]. This section will deal with noise and interference in multi-channel radio-relay systems.

Thermal noise voltages developed in the input stages of a receiver, either in a relay station or at the terminal station, will have an effect on the output signal-to-noise ratio which varies with the level of the incoming signal. When the incoming signal level is low, the a.g.c. action of the receiver will increase the gain of the receiver in an attempt to maintain the output voltage at a more or less constant level; unfortunately this means that the thermal noise generated in the input stage will be amplified to a greater extent. The other main source of

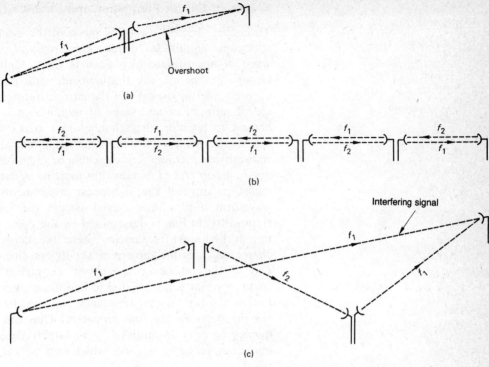

Fig. 11.8 Co-channel interference in a radio relay system

noise appearing at the output of the system is known as *intermodulation noise*. Intermodulation noise is produced by non-linearity in the amplitude/frequency and the group-delay frequency characteristics of the various parts of the system. Intermodulation noise has components at most frequencies and so sounds very much like thermal agitation noise. However, whereas thermal noise is continually produced intermodulation noise increases with increase in the signal level.

Adjacent-channel interference can occur on routes where all the available carrier frequencies are in use and therefore adjacent in-frequency carriers must be employed. Clearly, this form of interference can be minimized by the use of receivers of adequate selectivity. Co-channel interference can also exist if signals proper to one receive aerial are able to overshoot and be picked up by another aerial further along the route (see Fig. 11.8a). To reduce co-channel interference arising from this effect, two frequencies f_1 and f_2 can be allocated for use as carriers and alternate relay stations can use the same frequency (Fig. 11.8b). To reduce still further this form of interference, the path followed by the route can be zig-zagged in the manner shown in Fig. 11.8c; however the possibility of co-channel interference is not eliminated, as shown.

Choice of Carrier Frequency and Bandwidth

To enable the best use to be made of the available frequency spectrum, amplitude or frequency modulation of a carrier wave of appropriate frequency is used. Multi-channel telephony systems employ frequency-division multiplex and for each channel in the system the carrier frequency, or frequencies if more than one stage of modulation is used, must be chosen to position that channel in a particular part of the system bandwidth. To obtain the maximum utilization of the transmission medium (coaxial cable or radio relay system), the audio, group and r.f. bandwidths must be as narrow as possible whilst passing all the significant components of the signal waveform. For a line coaxial system the lowest frequency transmitted to line is determined by the need to avoid operation at the lowest frequencies where the attenuation/frequency characteristic of the cable is markedly non-linear. At the other end of the frequency spectrum, the higher the maximum frequency that is transmitted to line the greater is the attenuation of the line at that frequency and the closer must be the spacing between the line repeaters. Thus the maximum frequency to be transmitted is very largely determined by the minimum repeater spacing which can be economically justified.

To minimize the number of modulators and demodulators and the associated circuitry needed, a coaxial telephony system is assembled by suitably combining a number of C.C.I.T.T. 12-channel groups. The filters used to separate the audio channels are required to attenuate signals which are more than 600 Hz outside the 3.1 kHz channel bandwidth by at least 70 dB. To achieve such a high order of selectivity *crystal filters* [TSII] must be used. When the carrier frequencies for the 12-channel groups were originally chosen, crystal filters were only economically available at frequencies in excess of about 60 kHz. For this reason the channel carrier frequencies start at 64 kHz for channel 12 and increase in 4 kHz steps for each channel up to 108 kHz for channel 1. The lower sidebands for each channel are selected and so the bandwidth occupied by a group is 60.6–107.7 kHz, i.e. approximately 60–108 kHz.

Larger capacity systems are built up by suitably combining 12-channel groups. The ways in which the 12-channel groups can be combined have been specified by the C.C.I.T.T. and one example has been given earlier in this chapter (Fig. 11.1).

Multi-channel telegraphy systems are capable of transmitting 24 channels over a commercial quality speech circuit. The choice of carrier frequencies for the individual channels is primarily determined by the need to minimize inter-channel interference. Any non-linearity in the system will generate harmonic and intermodulation products, the most important of which, since they are of the greatest magnitude, are $2f_1$, $2f_2$, and $f_1 \pm f_2$ where f_1 and f_2 are two carrier frequencies. To minimize interchannel interference, the channel carrier frequencies are chosen to be the odd harmonics of 60 Hz, starting with the seventh. This means that the channel 1 carrier frequency is 420 Hz, channel 2 carrier frequency is 540 Hz, and so on up to channel 24 whose carrier frequency is 3180 Hz. The channel bandwidth is 50 Hz [TSII].

Wideband radio-relay systems must operate in the u.h.f. and s.h.f. bands because of the very wide bandwidths they occupy. Various frequency bands, listed in Table 11.2, have been allocated for this purpose by the I.T.U. (International Telecommunication Union). The C.C.I.R. issue recommendations regarding the division of each frequency band into a number of r.f. channels and, for example, Fig. 11.9 gives the recommendations for the 5925–6425 MHz band. The 500 MHz bandwidth is divided into a low and a high group of frequen-

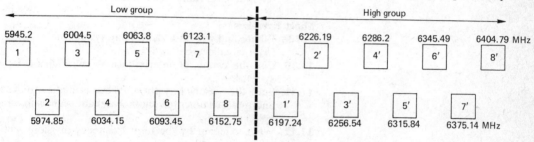

Fig. 11.9 Frequency allocation of a radio relay system

cies. At any particular relay station, all the transmitting channels are given carrier frequencies in one group and all the receiving channels have carrier frequencies in the other group. For example, at one station r.f. channels may be transmitted on low-group frequencies and received on carrier frequencies in the high group. Also, as shown in Fig. 11.8b, an r.f. channel received on a low-group carrier frequency at a relay station will be transmitted on the corresponding high-group carrier frequency, e.g. a channel received at a carrier frequency of 6004.5 MHz would be re-transmitted at 6256.54 MHz.

Exercises

11.1. Give an outline description of a typical microwave radio-relay system for carrying television and multi-channel telephony signals. Discuss the factors that would influence the choice of radio frequencies to be used for the system.

11.2. Discuss the sources of noise and interference in a multi-channel radio-relay system.

11.3. (a) Why is frequency modulation used in radio relay links in preference to amplitude modulation? (b) Why is pre-emphasis used for (i) multi-channel telephony and (ii) television signals? (c) Draw the pre-emphasis characteristic used in each case.

11.4. Draw the block diagrams of (a) the transmitter and (b) the receiver of an s.h.f. radio-relay link.

11.5. Describe, with the aid of diagrams, a multi-channel telephony system for use with a submarine coaxial cable.

11.6. Describe, with the aid of diagrams, a multi-channel telephony system for use in the inland telephone network.

11.7. Explain, with the aid of diagrams, what is meant by the terms supergroup and hypergroup. How may a multi-channel telephony system be transmitted over a s.h.f. radio-relay system?

11.8. Draw, and explain, the block schematic diagram of the repeater in a s.h.f. radio-relay system.

Short Exercises

11.9. Where and why in a radio relay system might (a) an isolator, (b) a circulator be used?

11.10. List the causes of interference in multi-channel radio-relay systems.

11.11. State the reasons why pre-emphasis is applied to the television and multi-channel telephone baseband signals in a radio relay system

11.12. What is meant by the term frequency-division multiplex and why is it used?

Numerical Answers to Exercises

1.2 41.67 W   **1.6** 360 V, 180 V

1.8 0.6325, 1.2 mW   **1.9** 0.33

1.11 (i) 990 kHz, 1000 kHz, 1010 kHz, (ii) 990 kHz, 1010 kHz, (iii) 990 kHz or 1010 kHz

1.13 144 kHz   **1.14** 136 kHz   **1.15** 0.57   **1.17** 3.147 V

2.1 70 kHz   **2.3** 8, 180 kHz   **2.5** 8.18%

2.7 46 kHz   **2.8** 66 kHz, 36 kHz, 6 dB down

2.15 Many possible answers   **2.19** 20 kHz

3.23 2045 Hz

4.3 1.2   **4.4** $L = 312$ nH, $C = 55.55$ pF

4.5 0.024   **4.8** (c) 5.657 dB assuming $G = 0$

4.11 5 V   **4.12** 0.375

5.13 26 dB   **5.14** 25%, 100 W

5.18 125 MHz,   **5.23** 1.291 m, 1.291 km

6.11 2.4 MW

7.1 408.33 W, 408.33 W   **7.2** 750 W, 300 W, 77.5%

7.8 3496 W, 14 124 W   **7.11** 0.833 A

7.14 1.57 A, $360°/\theta$ A   **7.15** 214.3 W

8.9 7.07 MHz   **8.11** $D = 27$, $D = 1.5$ **8.12** 6.842

9.1 40 pF, 195.5 μH, 283.3 pF, 140.7 μH

9.3 0, 0, 4 kHz   **9.4** 92.6 pF   **9.5** 3.33 kHz

9.13 40 pF, 232.6 μH, 258.3 pF

9.14 150.7 MHz, 129.3 MHz

Radio Systems III: Learning Objectives (TEC)

Index